Systems Engineering Management
Achieving Total Quality

James A. Lacy

Notices

IBM is a registered trademark of International Business Machines Corp.

Joint Application Design is a registered trademark of International Business Machines Corp.

National Electrical Code is a registered trademark of the National Fire Protection Association.

Post-it is a registered trademark of the 3M Corp.

SECOND PRINTING

© 1992 , 1994 by **James A. Lacy** All rights reserved.

1992 First Printing published by McGraw-Hill, Inc.

Printed in the United States of America. Except as permitted under the United States Copyright Act of 1976, no part of this publication may be reproduced or distributed in any form or by any means, or stored in a data base or retrieval system, without the prior written permission of the author.

Information contained in this publication has been obtained from sources believed to be reliable. However, neither the publisher nor the author guarantees the accuracy or completeness of any information published herein and neither the publisher nor the author shall be responsible for any errors, omissions, or damages arising out of use of this information. This work is sold or transferred with the understanding that the publisher and the author are supplying information but are not attempting to render engineering or other professional services. If such services are required, the assistance of an appropriate competent professional should be sought.

Library of Congress Cataloging-in-Publication Data

Lacy, James A.
  Systems engineering management : achieving total quality / James A. Lacy.
    p.   cm.
  Includes index.
  ISBN 0-9644627-0-2
    1. Systems engineering—Management. 2. Production management--Quality control. I. Title.
  TA168.L26 1991
  658.5'62—dc20   91-21576
                                                CIP

Contents

Preface viii
Introduction ix

Part I
The systems engineering process

1 Facets of systems engineering 3
  Defining systems engineering 3
  Facets of systems engineering 4
  Summary 7

2 Systems engineering activities 9
  Systems engineering process constraints 9
  Dimensions of systems engineering 10
  Overview of the logic dimension 10
  Problem definition 12
  Value system design 22
  Function analysis 26
  Systems synthesis 31
  Systems analysis 31
  Decomposition 42
  Description 45
  Handling changes 46
  Summary 47
  Further reading 48

3 Verification — 50
Defining verification 51
Verifying the product 51
Test and evaluation 54
Reducing integration and test time 55
Summary 55

Part II
Specialty engineering

4 Life cycle cost — 59
Defining life cycle cost 59
Benefiting from life cycle cost 63
Tasking the life cycle cost specialist 64
Controlling life cycle cost 65
Summary 66
Further reading 66

5 Supportability — 67
Defining supportability 72
Benefiting from supportability 72
Tasking the supportability specialist 72
Controlling supportability 73
Summary 74
Further reading 75

6 Reliability — 76
Defining reliability 76
Benefiting from reliability 78
Setting requirements 80
Tasking the reliability specialist 82
Tasking design engineers 88
Controlling reliability 91
Summary 93
Further reading 94

7 Maintainability — 95
Defining maintainability 95
Benefiting from maintainability 96
Setting requirements 97
Tasking the maintainability specialist 101
Tasking design engineers 106
Controlling maintainability 108

Summary 109
Further reading 110

8 Human engineering — 111
Defining human engineering 111
Benefiting from human engineering 112
Setting requirements 112
Tasking the human engineering specialist 114
Controlling human engineering 118
Summary 118
Further reading 119

9 Safety — 120
Defining safety 120
Benefiting from safety 122
Setting requirements 122
Tasking the safety specialist 125
Controlling safety 130
Summary 130
Further reading 130

10 Electromagnetic compatibility — 131
Defining EMC 131
Benefiting from EMC 133
Setting requirements 134
Tasking the EMC specialist 134
Tasking design engineers 138
Controlling EMC 139
Summary 145
Further reading 146

11 Testability — 147
Defining testability 147
Benefiting from testability 148
Setting requirements 149
Tasking the testability specialist 150
Tasking design engineers 153
Controlling testability 154
Summary 155
Further reading 156

12 Software — 157
Requirements 157
Technical planning 160

Technical control　162
　　　Summary　164
　　　Further reading　164

13　Producibility and manufacturability　165
　　　Defining producibility　165
　　　Benefiting from producibility　167
　　　Implementing producibility　168
　　　Summary　172
　　　Further reading　172

14　Value analysis　173
　　　Defining value analysis　173
　　　Benefiting from value analysis　174
　　　Implementing value analysis　174
　　　Summary　176
　　　Further reading　176

15　Design to cost　170
　　　Defining design to cost　177
　　　Benefiting from design to cost　177
　　　Implementing design to cost　178
　　　Summary　179
　　　Further reading　179

Part III
Management of systems engineering

16　Technical control　183
　　　Specifications　183
　　　Interfaces　189
　　　Statement of work　190
　　　Configuration management　192
　　　Technical performance measurement　194
　　　Managing risk　195
　　　Technical reviews　194
　　　Auditing systems engineering　195
　　　Summary　205
　　　Further reading　206

17　Work breakdown structure　208
　　　Defining terms　208

Benefiting from a WBS   210
Developing a WBS   211
Making revisions   219
Summary   219
Further reading   220

18 Cost and schedule control   221
Control techniques   221
Schedules   222
Systems engineering's affect on cost control   226
Earned value   227
Microcomputer tools   229
Summary   230
Further reading   230

19 The system engineering management plan   232
Overview   232
Planning in general   233
Using the plan   234
Applying MIL-STD-499A   237
Summary   242
Further reading   242

20 Proposals   243
Training   243
Writing proposals   244
Evaluating proposals   246
Summary   249
Further reading   249

Appendix A   MIL-STD-499A   250

Appendix B   MIL-STD-1521B excerpts   268

Index   320

About the author   326

To Barbara

Preface

This book was written because too many products are late to market, their development is costly, and they don't delight the customer. Systems engineering is a way to minimize these losses to society. For many companies and organizations, the ability to produce products effectively is gaining acute attention.

Clients, colleagues, and students have asked repeatedly for a guidebook to systems engineering. They desire a practical guide that is useful in everyday work. They want immediate, useful results in meeting deadlines and budget constraints.

I hope this book answers those needs of working engineers. It is designed for the world of work—with its chaos of real problems. For many people, ringing telephones and tasks that were due yesterday make orderly work a fantasy. Problems are difficult to recognize and define much less solve. This book can reduce your stress by helping you decide what problems are important and what to do about them.

Finally, I would like to dedicate this book to you, in appreciation of the way that engineers benefit other peoples' lives. I hope you also share the joy of making things work.

Introduction

This book will help you design and make products effectively and efficiently. Effectively means meeting the customer's expectations. Efficiently means providing a lower-cost product in a shorter time. Applying engineering efforts to transforming customer needs into a description of product performance and its physical realization is called *systems engineering*.

Companies and organizations are recognizing that systems engineering is crucial to customer satisfaction. Some symptoms of poor systems engineering are projects or products:

- Behind schedule, too late to markets.
- Over budget.
- Not able to meet technical expectations.
- Not able to meet customer expectations.
- Starting from single-point solutions.

If you have experienced any of these symptoms at your business, this book can help you and your organization. *Systems Engineering Management* is written for technical professionals leading projects and products. You might already be a systems engineer looking for ways to improve your professional skills. You might be looking for a promotion into the systems engineering field, or taking a course in a masters program. You might be a manager searching for ways to make your business more successful. This book is designed specifically to help you meet the many challenges of systems engineering.

There are three parts to this book. Part I describes the systems engineering activities. The logic dimension is emphasized, because the information dimension depends on your specific product. Part II describes specialty engineering tasks. These efforts are needed throughout the product cycle. Part III describes basic technical planning and control. Systems engineering management must be tailored to the specific situation. You'll learn how to:

- Transform customer requirements to "design to" requirements to "build to" requirements using a logical process.

- Establish planning and control to save time and money.
- Understand risk to prevent cost and schedule problems.
- Understand specialty engineering tasks and integrate them with the requirements.
- Use new quality tools for design engineering.
- Ensure that performance is consistent with customer needs to delight the customer for increased sales.

Also covered are Quality Function Deployment, Taguchi methods, Pugh convergence, and Joint Application Design. American systems engineering and Japanese quality techniques are a potent combination.

In this book, product and system have the same meaning. This is a convenience because the assumption is that you are building something for a customer. It also simplifies the focus by excluding lots of systems that are not products.

You might wonder about the widespread use of military standards in this book. This book is intended for both commercial and defense engineers. In no way is mandatory use of the standards suggested. Certain standards hold an accumulated body of knowledge built on years of experience. They have been proven on small and large systems. Take what makes sense, throw the rest away, and tailor it to your needs.

How do you best use this book? First, by tailoring the information to your particular work context. Second, by understanding systems engineering as a quality process, not a paperwork process. Third, by creating your own specific checklists, guidelines, and processes using this book as a starting point.

No book can tell you how to create a systems engineering management plan unique to your work context. Plans and processes must be tailored to fit the situation and problems. You should ask yourself the questions, "Does this make sense? Will the people doing the work follow the control systems I put in place? Is the information and quality derived worth the effort of feeding the control system?"

You will not find a separate chapter called "Quality". Every chapter is about quality. What is quality, anyway? The experts speak:

> Philip Crosby—"Conformance to requirements."

Dr. W. Edwards Deming—"Quality control does not mean achieving perfection. It means the efficient production of the quality that the market expects."

Dr. Armand V. Feigenbaum—"The total composite product and service characteristics of marketing, engineering, manufacturing, and maintenance through which the product and service in use will meet the expectations of the customer."

> Dr. Joseph M. Juran—"Fitness for use."

In a sense, quality is how a business or organization is managed. It is now a major criterion in any customer decision, which makes quality key to success and company growth. Systems engineering is an integral part of quality management. One of the major themes of systems engineering is "requirements before design." It is the intent of this book to show you how to achieve quality through systems engineering.

Part I

The systems engineering process

```
The Systems Engineering Process
  ├─ Facets of Systems Engineering
  ├─ Systems Engineering Activities
  └─ Verification
```

The systems engineering process is the process that you define to engineer products. In this section, systems engineering activities and logic are presented. These are the building blocks for a systems engineering process. The third chapter describes the verification of requirements.

1
Facets of systems engineering

Systems engineering is multidimensional, and even systems engineers have difficulty in defining it among themselves. It is important to take stock of the breadth of systems engineering before attempting to define a process. Also, for systems engineering to contribute to quality, its objectives should support the organization's objectives. Without exploring these areas first, you risk creating an ineffective systems engineering role. In this chapter, you'll learn the:

- Basic definitions related to systems engineering.
- Basic objectives of systems engineering.
- Facets of systems engineering.

Defining systems engineering

What is systems engineering? A one-sentence definition really does not do justice. It can be said that systems engineering is applying engineering efforts to customer needs and transforming them into a description of product performance and physical realization. It is also desirable that it be an optimal system definition and design that minimizes loss to society. Systems engineering integrates the total engineering effort to meet cost, schedule, and technical performance objectives. Systems engineering is both a technical process and a management process.

Some terms in systems engineering that merit further explanation include:

System—A set or arrangement of things that are related and form an organic whole. Relationships tie the system together.

Engineering—Putting scientific knowledge to practical uses.

System engineering—Systems engineering applied to a specific system or product.

Environment—All objects outside the system whose change in properties affects the system or whose properties are changed by the behavior of the system.

Systems engineer—One who does systems engineering.

System engineer—One who does systems engineering for a specific system.

Requirement—Something needed.

Process—A set of interrelated work activities characterized by a set of specific inputs, tasks, knowledge, and procedures that produce a set of specific outputs.

Project—A complex group of tasks performed in a definable time period to meet a specific set of objectives. It is usually unique and performed in less than three years under a limited budget.

Program—A long-term undertaking made up of one or more projects.

Facets of systems engineering

Systems engineering has many facets, including problem solving, decision making, risk, product complexity, and design. Emphasis on each facet and the activities supporting them vary by project, product, and company. Because of this variation, the total process will be unique for the specific project. Systems engineering is usually done within a project, because projects deal with new and unique things. For systems engineering to be useful, it must support the goals and objectives of the project. Looking at the project success factors gives clues as to what systems engineering should be about.

Project success factors

A project is generally considered successful if it:

- Meets customer needs and expectations and is accepted.
- Achieves its goals, including technical effectiveness.
- Is on schedule.
- Is on budget.

Contributors to successful projects are:

- A project team's knowledge and skills.
- Clearly established criteria.
- Properly defined requirements and deliverables.
- Accurate estimates of required resources.
- Proper resources to accomplish the project.
- Effectiveness of planning and control.
- Absence of bureaucracy and non-value-added tasks.

Project managers want an organized stack of clear goals, obvious performance measures, resources for every task, and time for rational, logical decision making. New managers are shocked to find reality far from management theory. Most of the workday is spent with other people in chaotic, sporadic contacts. Decisions are made under time constraints and are negotiated with many parties and groups. Goals are often conflicting, and without obvious means of measure. Standards established at the beginning of the project are often ignored or violated when the project is behind schedule. Besides running the business of the project, managers are often asked to act as the technical manager. Some relief is decidedly called for here. Part of the job of the systems engineer is taking technical responsibility for the project in support of the project manager.

Objectives of systems engineering

The main objectives of systems engineering are to:

- Minimize the loss to society.
- Minimize the time to market.
- Meet the customer's needs and expectations.
- Design a robust product.

Other objectives are to:

- Determine and derive requirements.
- Verify that requirements are met.
- Act as the user advocate.
- Ensure that the engineering effort is integrated.
- Ensure definition and design from a total-system approach.
- Achieve required effectiveness within risk, cost, and schedule considerations.
- Integrate requirements and specialty engineering efforts.
- Ensure compatibility of all system interfaces, including those within the system.
- Establish, control, and maintain an effective work breakdown structure.
- Evaluate the effects of changes on system performance, cost, risk, and schedule.
- Assess and manage technical risk.
- Document major technical decisions.

Role of the systems engineer

An effective systems engineer understands:

- Systems engineering.
- General engineering.
- Product-specific knowledge.
- Innovation and generating ideas.
- Working with other people.

An understanding of systems engineering alone does not make a systems engineer. There must be a foundation of engineering knowledge to build upon. Ignorance of technology and customer wishes for specific products make for poor requirements, which is poor systems engineering. Systems engineers need a broad background of experience. Innovation and ideas flow more easily once experience and information have percolated for a time.

Having a set of rules doesn't mean you will have good systems engineers. Much depends on the person. A systems engineer needs leadership to deal with change, because projects are all about change and newness. The project must be guided through uncharted territory. Systems engineering deals with complexity, and this calls for management skills, including working well with others.

Solving problems

One of the most important abilities that a systems engineer can have is the ability to solve problems. Recognizing, isolating, defining, and solving problems is essential. Formulating the wrong problems or stating them incorrectly wastes much time and resources.

Making decisions

Making decisions requires information for lower risk. The systems engineer is a broker and dealer of information. Systems engineers apply rules or logic to information in making decisions. Planning is really about making decisions. As you will see, the systems engineer spends a good deal of time planning.

Communicating

Projects cannot function without communications. The systems engineer sets up the internal technical communications paths within the project. Organizations and systems have a strong element of information flow and information theory. Without communications, the project is reduced to a group of people who only share the same work address.

Managing risk

Risk is potential loss. The systems engineer assesses and manages risk to minimize loss to society. Risk cannot be reduced completely, nor is it desirable to do so. Some risk must be present in a healthy organization. Much of the time, minimizing losses is the best that can be expected of the systems engineer. This is the nature of projects—that some territory is uncharted and holds surprises.

Complexity

Systems engineering grows in importance with complexity. Many companies discover the need for systems engineering about the time that their products evolve into complex systems. Systems engineering management is needed to deal with the complexity. Planning is done more thoroughly because shortfalls become more serious. Work must be chunked more carefully because so many people are involved, few of which have a total-system viewpoint.

Newness

As organizations move from precedented to unprecedented designs and products, the need for systems engineering increases, as well as the level of ability for leadership to handle change. When products are evolving in increments and building on previous models, information is more readily available to make decisions. Revolutionary or completely new products require systems engineering as a means to deal with newness.

Integrated approach

The approach to a system design considers several systems. Besides the product, or system being designed, the other systems worked concurrently are:

- Manufacturing.
- Support and repair.
- Facilities.
- Personnel.
- Procedural data.
- Computer support and programming.

The system (project) used to design these systems is part of the integrated approach.

Design

Systems engineering has a large component of design. Design is decision making and planning to satisfy specified needs. Design depends heavily on people and their particular filters for the way they process information. Filters include:

- Emotions
- Values
- Knowledge
- Habits
- Attitudes
- Experiences
- Intuition
- Imagination

Systems engineers need to be aware of their own filters as well as design engineers' filters to prevent unbiased communication. Communication is key to successful projects and customer satisfaction.

Summary

- The main objectives of systems engineering are to:
    - ~ Minimize the loss to society.

8 The systems engineering process

- ~ Minimize the time to market.
- ~ Meet the customer's needs and expectations.
- ~ Design a robust product.
- An effective systems engineer understands:
  - ~ Systems engineering.
  - ~ General engineering.
  - ~ Product-specific knowledge.
  - ~ Innovation and generating ideas.
  - ~ Working with other people.

2
Systems engineering activities

This chapter is called Systems Engineering Activities for a good reason. It cannot be called Systems Engineering Processes because the time element is not firmly defined. The activities are an important way of dealing with information. You, as the system engineer, are making decisions and solving problems based on the input information. Some of the activities provide a framework and discipline for creativity in imagining alternative solutions. When the activities are tailored to specific projects and properly time sequenced, then a total process can be defined. In this chapter, you'll learn the:

- Constraints on the systems engineering activities.
- Logic dimension of systems engineering.
- Activities of systems engineering.

Systems engineering process constraints

It would be nice to talk about a process that produces an ideal optimal product. Perfection is never achieved in real products, however. You must be aware of the constraints that prevent perfection. One of the first constraints is cost. Both engineering costs and the cost of the end product will force a halt at some point to engineering activities. A second constraint is the time to market, which will force a deadline on completion of activities. No responsible manager will let the process go forever. The process receives incomplete information to begin with, because knowledge increases as time goes on. These constraints drive systems engineering to be practical and less than perfect.

Defining the process is not easy because the process is not always sequential. The information needed does not always come at the desired time. New information can cause renewed decision making, which makes the process recursive, or occurring again or periodically. This is not unusual. Iterative means that the process is repeating or repetitious. Iteration happens because of human mistakes, poorly defined requirements, or an increase in knowledge. Excessive iteration, however, is a sign that the process or people are not working efficiently.

Dimensions of systems engineering

There are four dimensions of systems engineering[1]. They are:

1. Logic
2. Information
3. Creativity, innovation, invention
4. Time

The logic dimension refers to the activities done by systems engineering. This dimension has a flow of logic; not a flow of time. When logic is combined with time, activities result.

The information dimension attends to specialized information and knowledge from engineering, the customer, and the product area. Information is much better when it is correct. Knowledge implies understanding, but engineers do not always have full understanding nor is it always necessary. Many engineering feats have been accomplished by using what worked at the time. Engineers are not scientists who must have mathematical formulas. Rather, engineers make practical use of science.

The creativity dimension deals with ideas. Ideas come first for any activity or physical product. Innovation is a change to an existing idea or product. Innovation can result from combining two unrelated subject areas. Invention is something new or thought about for the first time. The creativity dimension is strong when generating alternative designs.

Table 2-1. Major Project Time Phases Examples.

Defense	Commercial
Statement of need	Marketing
Program initiation	Project planning
Concept exploration	Concept design
Concept selection	Detail design
Demonstration/validation	Pilot or prototype
Program go-ahead	
Full-scale development	System development
Production ratification	
Production	Production
Deployment/initial operating capability	Distribution
Operation and support	Operation
Disposal	Retirement

---

[1] A three-dimensional model was proposed by Arthur D. Hall in "Three-Dimensional Morphology of Systems Engineering" (see Further Reading at the end of this chapter). This work was expanded upon by J. Douglas Hill and John Warfield in "Unified Program Planning." Andrew Sage also describes the Hall model in his *Methodology for Large Scale Systems*. Systems engineers should be aware of these pioneering works, among others. This book builds on these works, with the addition of the creative dimension.

The time dimension describes the phase in which a project and product pass through. This time is bounded when in a project. Information increases with time. Time occurs in most systems engineering activities as a variable, such as the profile of use for the product.

Overview of the logic dimension

The general logic dimension of systems engineering includes:

- Problem definition
- Value system design
- Function analysis
- Systems synthesis
- System analysis
- Decomposition
- Description

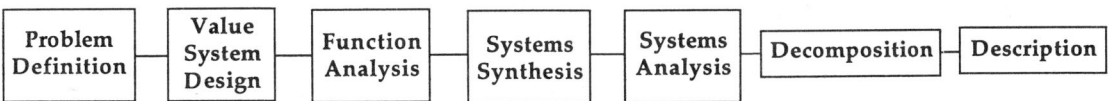

Fig. 2-1. Logic dimension overview.

Again, the flow is by logic, and not by time. To prevent confusion, we will call the things in the logic dimension *activities*. The word *process* is reserved here for the total systems engineering process. These definitions are arbitrary and only for exploring the multiple dimensions. In reality, there are numerous processes performed on a project. At this point, you can avoid the argument of defining "the" systems engineering process. You will have to define that for yourself. It now is important to deal with the flow of logic, first.

Another caveat must be stated before proceeding. The previous activities should not be considered "the x-step process." The top activities of the logic dimension hierarchy are seven, for reason of human cognition, and not for an iron-clad rule. As you will see, within each major task there are many subactivities, some of which are actually major activities by themselves. The number of activities is not important, only that logic is served in this dimension.

Problem definition occurs repeatedly during the project life cycle. Needs assessment and requirements are major themes. The systems engineer has a tremendous concern with requirements, for these are the products of engineering to manufacturing. Most of the systems engineer's work leads, in some way, to defining requirements.

Value system design sets objectives and measures. Systems engineering determines yardsticks to measure alternatives. Criteria, weighting factors, and utilities are all part of this task.

Functional analysis describes the required functionality. Functions occur in the product (prime system), the support system, the manufacturing system, and in all systems surrounding the product. The systems engineer is interested in the sequences and interrelations of the functions. Functions are a subset of requirements.

Systems Synthesis collects, innovates, and invents alternatives. The creative dimension ties strongly to this task. Alternatives can be different configurations of the same components. Likely solutions are brought together from the infinite set of solution possibilities.

Systems Analysis includes analysis, optimization, and decision making. Analysis means to separate. Systems engineering applies analysis to the alternatives using the constraints and criteria. An alternative is chosen that best meets needs. This alternative is optimized through further study and analysis. Systems engineering makes a final decision to accept or reject and baselines that decision. This is a likely point for recursive behavior.

Decomposition takes an alternative and decomposes, allocates, and defines interfaces. This task prepares for the next time phase, or next lower level hierarchy of the system. In effect, it is planning.

Description means documenting decisions, definitions, values, analyses, or any important information that might be needed in the future. Systems engineering seeks the minimum necessary documentation. Iteration will generate more documents than can be remembered and used.

Within each of these activities are numerous sub-activities. The order of performance of the sub-activities is also dependent on the information flow, in addition to logic needs.

Problem definition

Properly defining the problem helps immensely in defining solutions. This can be visualized as bounding the problem space so that a solution space can be mapped into it. Some of the sub-activities of the Problem Definition task are:

- Needs assessment.
- Constraints.
- Scenarios.
- Scope.
- Developing requirements.
- Identifying key people needed.
- Identifying the degree of need.
- Identifying alterables.
- Bounding and partitioning the problem.
- Environments.
- Describing interactions.
- Determining normal and out-of-normal conditions.

Needs Assessment

A needs assessment is done to find out what the customer's problems are. Without a defined problem, you probably will have no product to sell. The needs assessment

Systems engineering activities   13

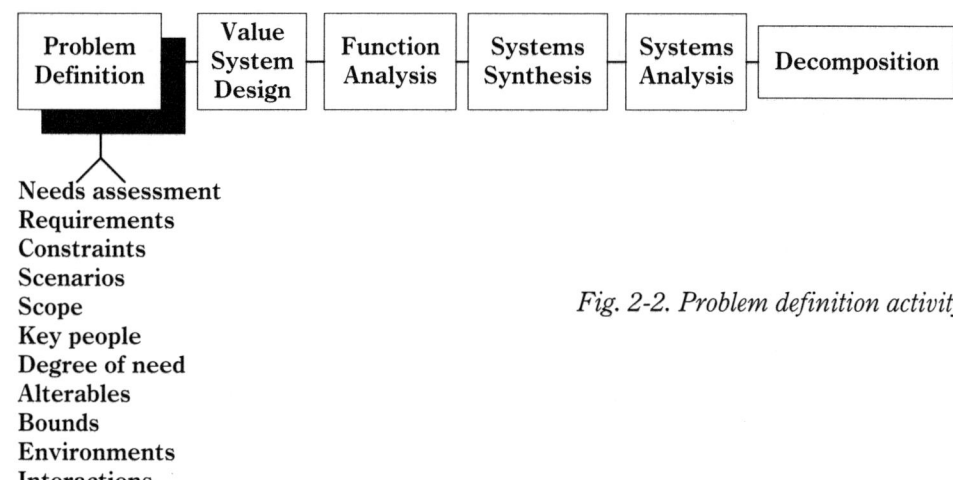

Needs assessment
Requirements
Constraints
Scenarios
Scope
Key people
Degree of need
Alterables
Bounds
Environments
Interactions
Normal and out of normal conditions

Fig. 2-2. Problem definition activity.

defines what is needed, not the solution. The needs assessment at the system or product level might include these questions:

- Who is the customer for this product?
- What does the product do for the customer?
- Why does the customer need this product?
- What can the customer afford?
- Where does the customer use the product? What are the environments?
- How will the product be maintained and supported?
- How much or often is the product used?
- What expectations does the customer have?
- When does the customer expect the product?
- What critical measures of effectiveness must be met?
- How will the customer use the product?
- Are the users different than the customer?
- Are there critical safety or liability issues?
- Are any operations time-critical?

Developing requirements

Requirements are needs. Requirements can be quantitative, qualitative, derived, or allocated. Quantitative requirements are measurable in terms of numbers, such as 600 miles per hour. Qualitative requirements are measurable, but in terms of yes or no. A requirement for black paint is verified by a yes, and it is painted black. Derived requirements are requirements transformed from a higher-level requirement. For example, a requirement for low fuel use may transform to a requirement for low weight. An allocated requirement is proportioned or divided from a higher-level requirement. For example, the system weighs 100 pounds, and box 1 weighs 70 of that, and box 2 weighs the remaining 30 pounds. Derived and allocated requirements result from requirements flowdown by systems engineering. Attributes of good requirements are:

- Easily verifiable.
- Unambiguous.
- Complete.
- What is needed, not how to do it.
- Consistent with other requirements.
- Proper for the level of system hierarchy.

Stated requirements that are not verifiable are not requirements. Using words such as resistant, sufficient, or excessive will result in unverifiable statements. There must be only one possible interpretation of the requirement for it to be unambiguous. Requirements must be complete enough to stand alone. The word "shall" means mandatory in a requirement.

Constraints

Constraints are different from requirements. Constraints are what you can't or won't do. Constraints include:

- Funding.
- The number and skill of people to solve the problem.
- The manufacturing system.
- Facilities.
- Time and schedule.
- Legal and statutory.
- Tools.
- Policies and procedures.
- Critical-resource usage.
- Political.
- Resource availability.
- Patents.
- Past habits, product series, and contracts.
- A components' characteristics.

Systems engineering identifies constraints that are used to eliminate possible alternatives. Constraints should be used to eliminate alternatives that are not feasible before using criteria to select the best alternative.

Scenarios

Scenarios include outlines and synopses of proposed events concerning a customer's problem. One of the most common descriptions is the operations concept. The operations concept is a time-sequence description of events and functions in the use of a product. The term *mission profile* is sometimes used to include both operations concept and environmental profile. The questions answered by the operations concept include:

- Why must these things happen?
- What is supposed to happen?
- Who or what is doing these functions or behaviors?
- When do these things happen, and in what order?

A single chart is too confining for comprehensive information. Several charts typically show the overall operations and the details for each major operation. The information is then available for derivation of further requirements.

Scope

In scoping, you decide the rough magnitude of the problem. You decide if this problem is a good match for your organization's capabilities.

Identifying key people

Key people are those people who are essential to the project's success. They have a technical or management ability that is identified as critically needed.

Identifying the degree of need

The degree of need tells you how serious the problem is. The deadlines for delivering the project also tells the degree of need.

Identifying alterables

Alterables are things that can be changed. They might include the environment, customer perceptions, maintenance concepts, or the method of production, for example.

Bounding and partitioning the problem

Bounds are the limits of the problem. Will the product be a part of a series of items? Will the product only be sold in the United States? How many products are projected for manufacture? Bounds also determine constraints as well as identify critical issues that must be addressed to achieve success. Bounds can sometimes be altered for a better choice of alternatives.

Environments

Environments include both natural and induced. Systems engineering should consider normal and out-of-normal environmental conditions. Examples of natural environments include:

- Freezing rain
- Fungus
- Hail

- Humidity
- Ice
- Lightning
- Pressure
- Solar radiation
- Rain
- Salt spray
- Snow
- Temperature
- Wind

Induced environments include:

- Acceleration
- Electromagnetism
- Electrostatic
- Shock
- Vibration

The environment of a system determines its design. The degree of fit to the environment and robustness to environmental changes are measures of good design. Therefore, systems engineering must define the environments and the system/environment interface.

One way to organize this information is through an environmental profile chart. One axis of the chart represents the time phases of interest, such as use, maintenance, shipping, and storage. The other axis represents environmental conditions, such as vibration, temperature, shock, and humidity. The quantitative figures of environments for each time phase are recorded in a two-dimensional chart, such as the one shown in FIG. 2-3.

Many project activities will use the chart and information in it. Mechanical structures must have the shock and vibration data to design. Reliability predictions and calculations depend on accurate temperature and vibration data. The duration of the conditions is as important as the numerical values. In short, all designers need this information before starting design.

Part of problem definition is determining the extent of operation and support by the user. Knowing who uses the product and when is important. The same applies to the maintenance of the product. A utilization concept must evolve that satisfies user needs and fits the product. A maintenance concept must be compatible with the intended use and the product itself.

Describing interactions

Interactions must be analyzed for both positive and negative effects. Requirements may be conflicting. Requirements and constraints may interact. Systems engineering seeks a workable solution in these conflicting elements.

Item	Flight Phase								
	Prelaunch		Initial Ascent		Transsolar Insertion	Solar Orbit Data Acquisition			
						Event			
	Precount and Countdown	Hold for Launch Window	Launch and Stage I	Stage II	Stage III	Zone I Collect Data	Zone II Collect Data	Zone III (Behind Sun) Collect Data	Zone IV Collect Data	Zone V Collect Data
Mission time event starts	...	...	0	5.9 min	9.2 min	2 hr	74 days	113 days	273 days	305 days
Duration of event	...	...	296 sec	200 sec	60 sec	74 days	39 days	160 days	32 days	74 days
Mean solar dist., AU	...	...				1.0 to 0.5	0.5 to 0.3 to 0.5	0.5 to 1.0 to 0.4	0.4 to 0.3 to 0.5	0.5 to 1.0
Mean Earth dist.	...	...	0 to 476,000 ft	584,000 to 550,000 ft	550,000 ft to insertion	10^{-4} to 0.5	0.5 to 1.5	1.5 to 2.0 to 1.4	1.4 to 0.6	0.6 to 0.2
Lateral shock (max) on payload, g in 10 ms	Free drop 1 in.	...	2.5	2.5	3.0	0	0	0	0	0
Ambient pressure, psia	14.7	14.7	14.7	0	0	0	0	0	0	0
Controlled temperature, °C	14° to 26°	14° to 26°	14° to 26°	14° to 26°	14° to 26°	14° to 26°	14° to 26°	14° to 26°	14° to 26°	14° to 26°
Acoustic, db	100	100	140	92	77	...	...	...	...	...
Vibration (payload) Sinusoidal, 20 to 2000 cps, g, rms	...	...	10	8	4	...	...	...	...	...
Random, 20 to 2000 cps, g²/cps	...	...	0.15	0.15	0.10	...	...	...	...	...
Longitudinal load, g	0	0	6.9	2.3	23	0	0	0	0	0
Lateral load, g	0	0	0	0	8	0	0	0	0	0
Thermal radiation, BTU/hr-ft²	...	...	...	540	540	...	4930 max	...	4930 max	...
Solar flare, protons	...	...	...	...	...	Spectrum follows modified NASA A,R,C, model. Flux of protons exceeding 120 MeV energy is 10⁹, and flux of those exceeding 0.75 MeV is 10¹³ particles/sq cm sec sr				
Solar wind, plasma flux, max	...	...	...	...	...	Flux of 1.2 × 10⁸ particles/sq cm sec; velocity of 500 km/sec; energy of 0.5 to 1.0 keV				
Solar E-M radiation, X-ray through infrared	...	...	...	...	...	Maximum X-ray flux during flares follows modified Johnson spectrum. Max energy is 10¹¹ keV/cm² sec keV for X-rays of energies up to 0.5 keV, and 10² keV/cm² sec keV for X-rays of energies up to 200 keV				
Pressure	...	...	...	...	...	Less than 10⁻¹⁰ mm of mercury				

Fig. 2-3. Matrix of environments and mission phases. (Source: NASA SP6503)

18   The systems engineering process

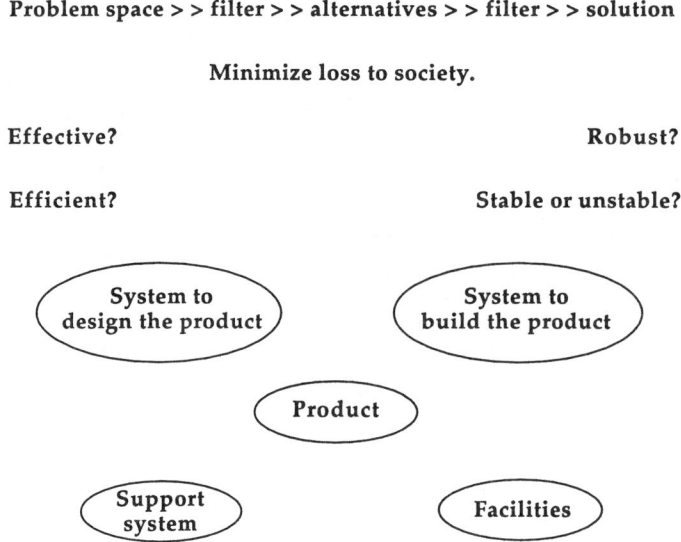

Fig. 2-4. Requirements of the product coexist with other systems.

Determining normal and unnormal conditions

The systems engineer must define both normal and out-of-normal conditions. These can include environments, for example. Users can sometimes find new and unusual ways to operate the product. Inputs to the product might fall out of the bounds set by the designers. Examples include the wrong data format or the wrong fuel. Designers need to know the normal and unnormal conditions for effective product design.

Requirements and constraints definition tools

Defining requirements and constraints can be difficult, and you might want to use some common tools to do this. Constraints are things you cannot, or won't, do. Two areas of constraints are patents and legislation.

The first step in avoiding patent infringement is to conduct a patent search. Patent searches reveal:

- Companies working in the same technology.
- Technology in expired patents that can be used for free.
- The prior state of the art in technology.
- Technology covered by patents that you must avoid, buy, license, or challenge.

Legislation is another source of constraints. Safety is an area well covered by regulatory requirements. Trade journal articles, professional consultants, and legal professionals are all tools you can use.

Defining requirements means finding out what the customer wants and organizing these wants. The tools for requirements include customer surveys, Joint Application

Design (JAD®), marketing information, charts, and Quality Function Deployment (QFD). The customer will also tell you many of the requirements, much of which can be learned through:

- Customer interviews.
- Customer questionnaire.
- Consumer panels.
- Product testing services.

Identified customers can be analyzed with demographics and psychographics, which can often predict customer preferences that might not be explicitly stated. When products are new, customers might not know what they want because they have never used the product before. In this case, there might not be a voice of the customer but a projection based on past preferences. A demographic and psychographic survey can be had through your marketing specialists, or through consultants. *American Demographics* magazine has numerous advertisers offering services in these areas as well.

Joint Application Design

Joint Application Design (JAD) is a common effort performed by the system users and system designers. It centers about a structured workshop called the JAD session. The workshop has a detailed agenda, a moderator/leader, and a scribe who records the agreed-upon requirements. The beauty is in the short time it takes to arrive at requirements, agreed to by the user/customer, and recorded real-time!

The JAD concept credit goes to Chuck Morris of IBM about 1977. In 1980, IBM Canada adapted and refined the tool. JADs have since fanned outside IBM through training courses and are now used for all kinds of applications, including the original management information systems. JAD tasks include:

- Project Definition:
    - Interviewing users.
    - Creating the participation list.
- Research:
    - Interviewing designers.
    - Learning about the system.
- Preparation:
    - Preparing the Working Document.
    - Preparing the session script.
    - Scheduling the meeting.
- JAD Session
- Final document:
    - Reviewing and updating the draft.
    - Getting signatures on the document.

Using neutral, trained moderators and scribes works best. The key is preparation. For the meeting to be focused, the designers must have a good idea of the requirements for which they are looking.

20 **The systems engineering process**

JAD sessions are an excellent way to converge diverse groups to an agreed specification or set of requirements. They can shorten the development time of a product dramatically by forcing all the key players into one room without disturbances.

Noncustomer interactive analysis

Not all requirements analysis is customer interactive. Other sources of requirements include:

- Literature research.
- Computerized databases.
- Trade journals.
- Trade shows.
- Market research.
- User characteristics databases (for example, anthropometric).
- Forecasting.
- Modeling.

Tools

Systems engineering uses the cross correlation chart, the self-interaction matrix, and the N × N chart[2]. The cross-correlation chart shows the relationship between two dimensions. The self-interaction matrix shows internal relationships within one dimension. The N × N chart shows interfaces and relationships.

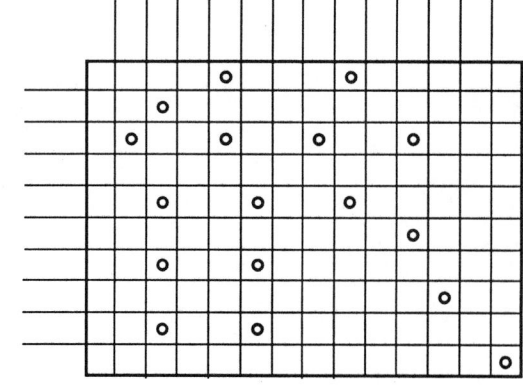

Fig. 2-5. Cross correlation chart layout.

---

[2]R. J. Lano developed the N-squared chart. It is usually used for data, process, or hardware interfaces. The basic idea is shown in FIG. 2-7. The functions or processes are placed on the diagonal. Information flows in a clockwise direction among functions. Feedback is shown in the lower, left-hand square intersecting the two functions. All outputs are horizontal, and all inputs vertical. Names of the functions and interfaces are listed in the squares.

Quality Function Deployment (QFD) is an excellent tool for both planning and requirements flowdown. It combines elements of the cross-correlation chart and the self-interaction matrix. You will read more about QFD in the Decomposition section of this chapter.

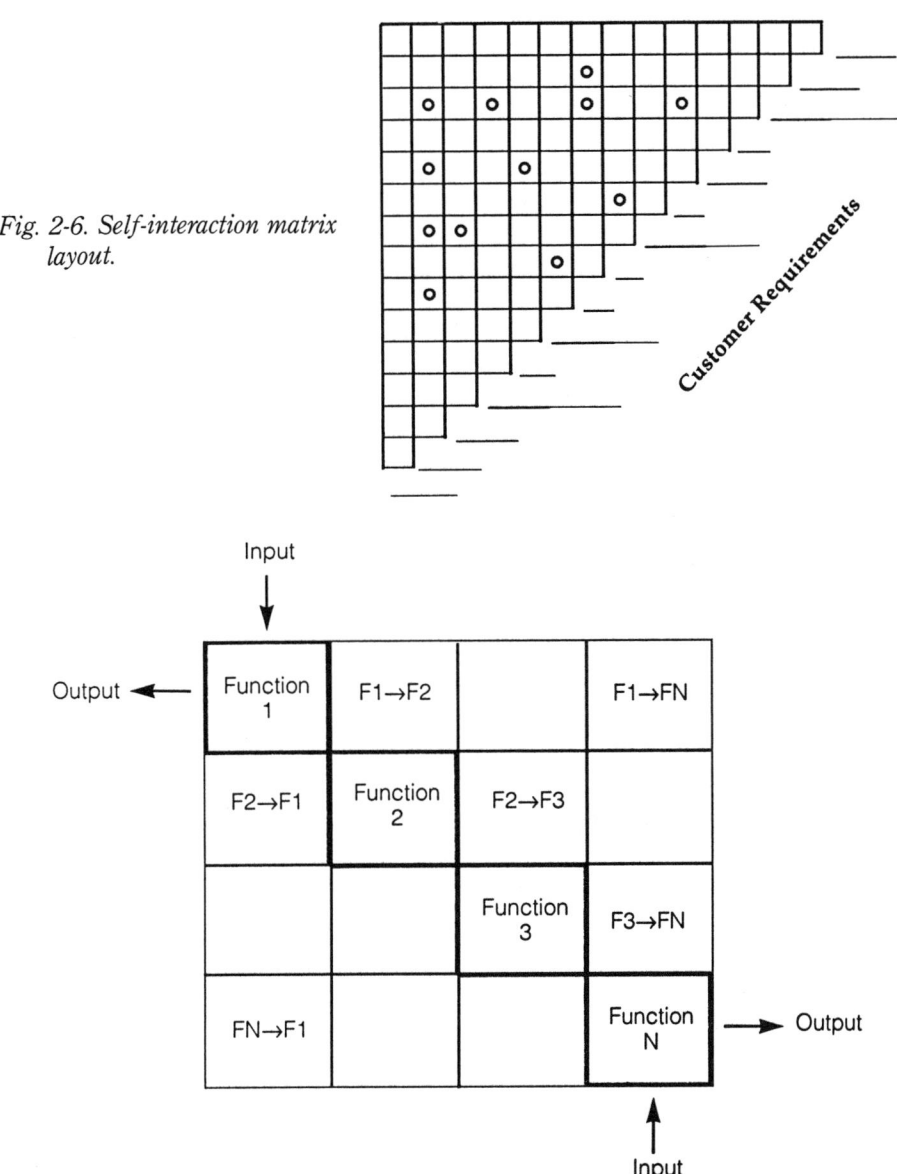

Fig. 2-6. Self-interaction matrix layout.

Fig. 2-7. N × N chart diagram.

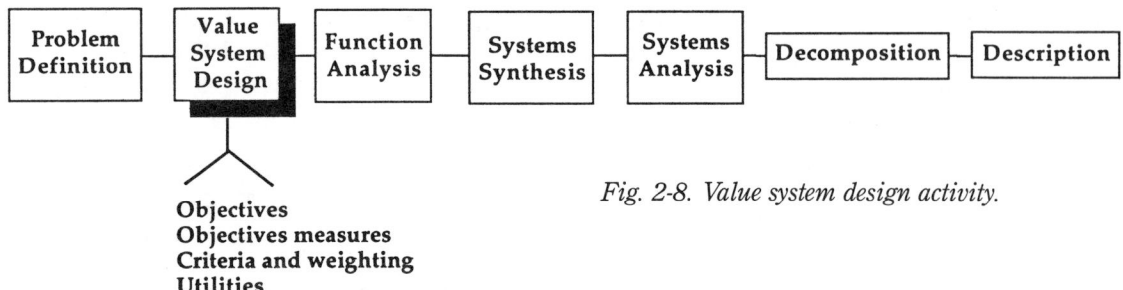

Fig. 2-8. Value system design activity.

Value system design

Value system design requires looking at four areas: objectives, objectives measures, criteria and weighting, and utilities. These define what is desired of the product.

Objectives

Choosing the right objectives is crucial for product success. Choosing the wrong objectives will certainly lead to the wrong solution. With the right objectives, you have a better chance of selecting the right solution even if it is less than optimal.

Setting objectives has strong elements of the creative dimension. Objectives should be stated from the view of what is needed and not how to do it. Objectives are often stated as maximization, minimization, or closest fit to target. The English language can be a hindrance with its ambiguities and slanted meanings. Therefore, be sure each objective is measurable.

Objectives must be consistent with higher-level objectives. Otherwise, efforts can be wasted on objectives that are not important.

Objectives measures

Objectives are sometimes stated as measures of effectiveness. The effectiveness of a product determines its "worth." Systems engineering seeks "worth" at minimum cost. A measure of effectiveness has these characteristics:

- Relates to performance.
- Measures quantitatively.
- Is simple to state.
- Is complete.
- Is a function of time.
- Is a function of the environment.
- May be statistical or measured as a probability.
- Is easy to measure.

For example, a measure of effectiveness for an automobile might be fuel consumption in miles per gallon under specified environmental conditions.

Effectiveness at a system level has several definitions. One definition has effectiveness made up of these factors:

- Performance.
- Availability—the probability that a product is ready for use when needed.
- Dependability—the probability that a product behaves reliably in use.
- Utilization—the actual use of the product versus its potential.

Measures of effectiveness have many factors. To help you identify critical factors that contribute to product effectiveness you might want to show these graphically as a performance hierarchy tree, for example. Make measures of effectiveness quantitative expressions. Analyze supporting measures of performance from the measures of effectiveness. Make the measures of performance specific, and derive lower-level measures from these. A complete hierarchal structure formed this way shows the critical technical performance measures.

Criteria

Criteria are different from constraints. Constraints are the "musts," the restrictions, the limitations that must be met. Constraints can be used as screening devices to filter out alternatives. Once the alternatives have been screened, constraints cannot help determine the best alternative. Constraints are pass or fail.

Criteria are continuous. Criteria provide a way of judging feasible alternatives. For example, a criterion might be lowest cost, most range, fastest acceleration, or closest flow rate to 10 gallons per minute.

Sometimes, a measure can be used as both a constraint and a criterion. For example, as a constraint, the product must cost no more than $10,000. The customer prefers the lowest cost below that (a criterion). Cost is the measure here for both particular constraint and criterion. Criteria come from:

- The customer.
- Quality Function Deployment charts.
- Functions or behaviors.
- Measures of effectiveness.
- Measures of performance.
- Cost.
- Schedule.
- Manufacturing.
- Product support.
- Project and organization objectives.
- Other considerations of importance.

Weighting

Criteria do not have equal importance. Weighting factors are assigned to criterion to indicate importance. In selecting alternatives, criteria weighting seeks a closer problem to solution match.

Weighting can be determined empirically or subjectively. The empirical method derives weights by determining how much of each elementary measure contributes to a general outcome measure. Large numbers of measures require statistical analysis. The scenarios and environments for the studies must be chosen carefully. The measures of success or stated customer desires will drive the weight factors.

Subjective weighting uses the judgment of experts. One widely used method gives raters a fixed number of points, 100 or 1000, to allocate the criteria. The distribution of points reveals its importance. Another technique is using experts to score existing alternatives, and then deriving the criteria and weighting factors from the preferred alternatives.

You might be aware of some of the concerns with the weighting methods. The empirical techniques are sensitive to the specific conditions for which they were measured. The subjective techniques depend on the judgment of the experts. New products might not have strongly identified criteria. Depending on the rating method as absolute ignores the inherent uncertainty. Scoring should always be challenged, and some recursion can occur. Figure 2-9 is an example of a scoring method.

		Alternatives					
		1		2		3	
Criteria	Wt	Score	Wt Score	Score	Wt Score	Score	Wt Score
Cost	40	3	120	4	160	5	200
Reliability	10	2	20	3	30	3	30
Maintainability	5	1	5	4	20	3	15
Easy to manufacture	5	2	10	3	15	4	20
Easy to use	5	5	25	4	20	4	20
Product safety	3	4	12	5	15	5	15
Easy to test	2	3	6	3	6	2	4
Performance	30	3	90	4	120	5	150
Total	100		288		386		454

Fig. 2-9. Scoring method example.

Utilities

Utilities describe the relative value of a criterion for different levels of performance. Utility curves can be used in the scoring process and for sensitivity analysis. Utility curves are graphs of a characteristic versus its relative numeric value. Examples of utility curves are shown in FIG. 2-10.

Calculating loss is one way to plot utility curves. In FIG. 2-10, missing the schedule results in a total loss. Loss increases as the meantime-between-failures decreases. Estimating the loss at intervals results in points that can be graphed. These graphs show sensitivities in easily understandable form.

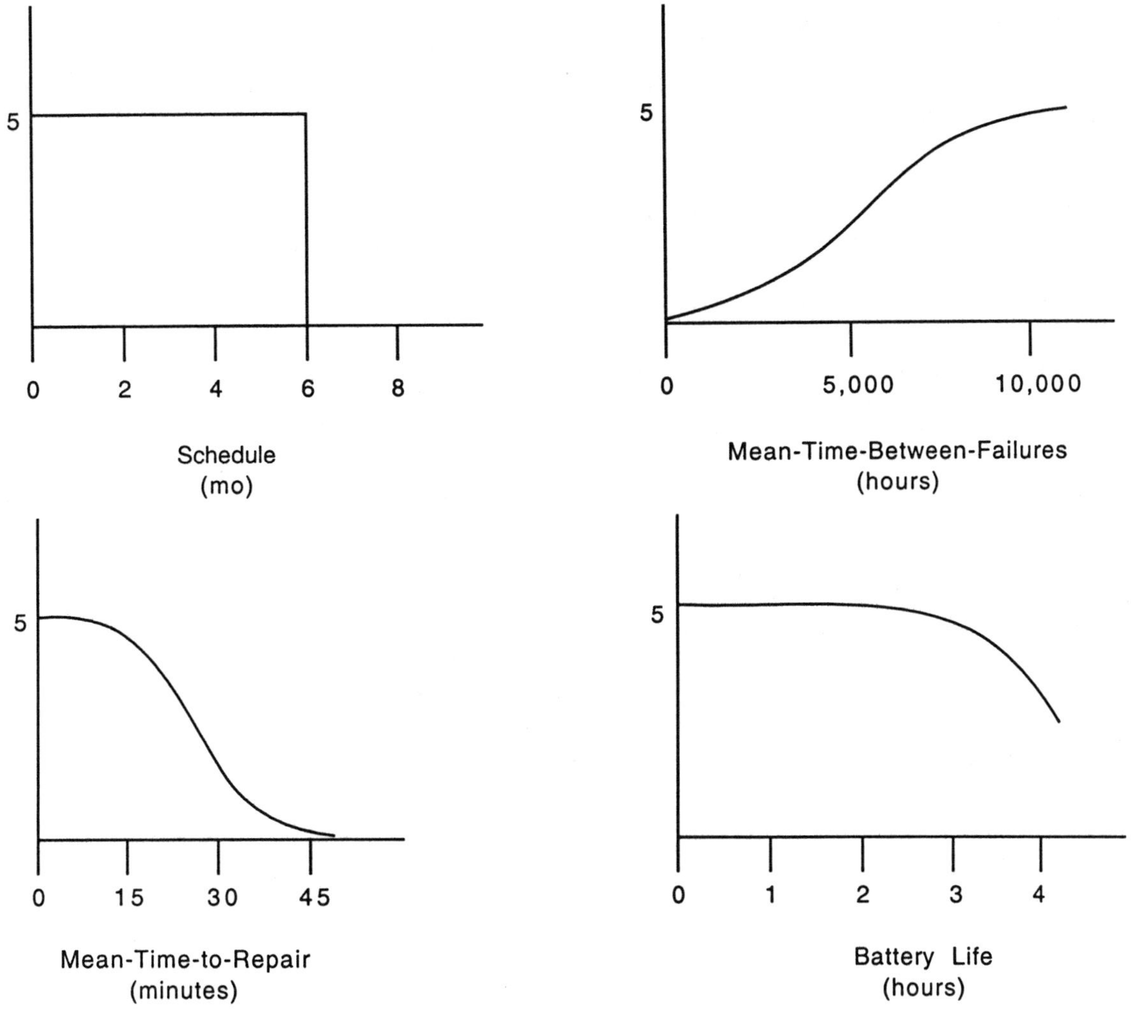

Systems engineering activities   25

Fig. 2-10. Examples of utility curves.

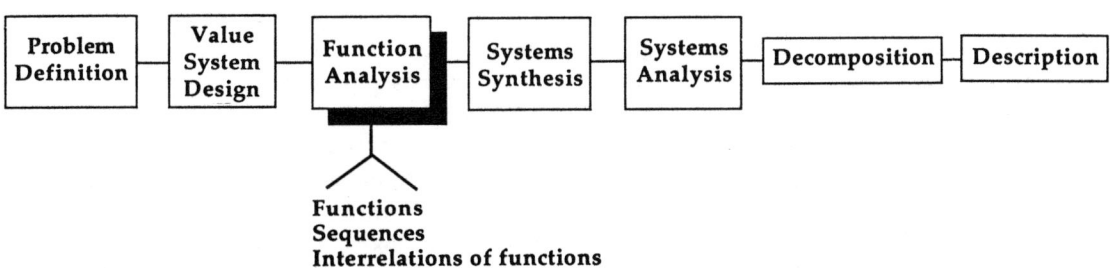

Fig. 2-11. Function analysis activity.

Function analysis

Function analysis has two extremely important benefits. It discourages single-point solutions, and describes the desired behaviors that become requirements.

A design team typically has experts in the product field. This knowledge makes for a better design. Unfortunately, this also introduces a bias about what will and won't work. Expert designers often drive towards single-point solutions without examining alternatives. Function analysis yields a description of behaviors rather than a parts list. It shifts the viewpoint from the single-point physical to the unconstrained solution set.

Products have desired behaviors. These may be behaviors that the customer sees, or behaviors internal to the product. Behaviors or functions interest systems engineering because they are really requirements. You want to know the functions, their sequences, the interrelations, and critical timing to derive requirements.

Function analysis also allows better functional and physical groupings for interfaces. Verification, testability, and maintainability improve because of function and interface analysis. Minimize the inputs and outputs for a subsystem, and minimize the interactions between subsystems.

Identifying functions

There are a large number of possible functions and functional sequences for a product. The selection of functions drives the later requirements and design. Functions are usually identified from higher-level requirements, desired behaviors, or higher-level function analysis.

Functions are expressed in two words, as an active verb and as a noun. Ideally, the noun should be a measurable attribute, and the verb-noun combination something verifiable. Nouns must not be a part or activity. Systems engineering keeps the viewpoint of the user. Functions are expressed as what the customer desires. This can prove difficult at first. For example, "provide power" is better stated as "power electronics." Active verbs are something that can be demonstrated. Keep it functional, and avoid describing physical parts.

Most engineers have no difficulty identifying primary or active functions of the product. Supporting functions seem to be harder to grasp. Supporting functions ensure reliability, make the product easy to use, delight the customer, or please the senses, for example.

You should realize that there is not "a" function analysis. Instead, a series of function approaches can be analyzed at a given level of product hierarchy. Each level of hierarchy within the system or product requires function analysis supporting the next higher level.

Systems engineering studies analyzes alternative function approaches for performance requirements, constraints, and areas of risk. Inputs, outputs, environments, time constraints, and other considerations are documented for functions. Similar functions are referenced for possible common solutions. Systems engineering must treat function analysis seriously for better requirements definition.

Some systems engineers consider functions important enough to assign function champions. These people are responsible for their assigned functions. They make sure the functions happen in the product.

Systems engineering activities 27

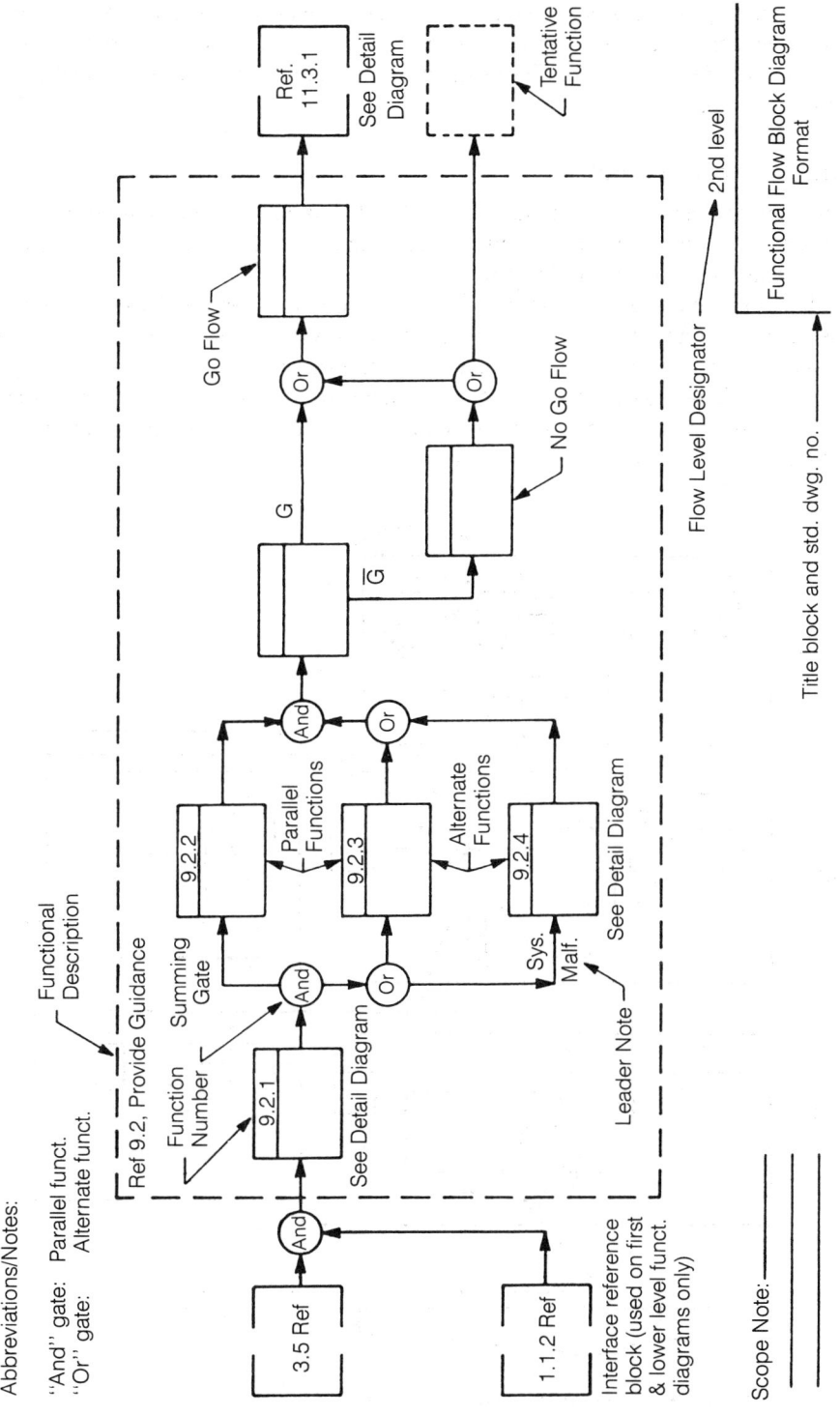

Fig. 2-12. *The general format and symbols for a functional flow block diagram.* (Source: FM 770-78)

28   The systems engineering process

The prime product is not the only place where functions are analyzed. The manufacturing system, support equipment, people, facilities, and other related systems or elements have functions also. These functions depend on the product functions and coexist with them.

Tools for function analysis

No function analysis tool is sufficient by itself. Information is handled differently by the various tools. There are a few common tools that are widely used, however, one of which is the functional block diagram.

The functional block diagram identifies functional areas and their information paths. This diagram can be decomposed into functions at lower levels. Figures 2-12, 2-13, and 2-14 are examples of functional block diagrams.

The functional flow diagram is a time-sequenced diagram showing the functions to be performed. The emphasis is on the sequence of functions. Interfaces and information flow are not shown on this diagram.

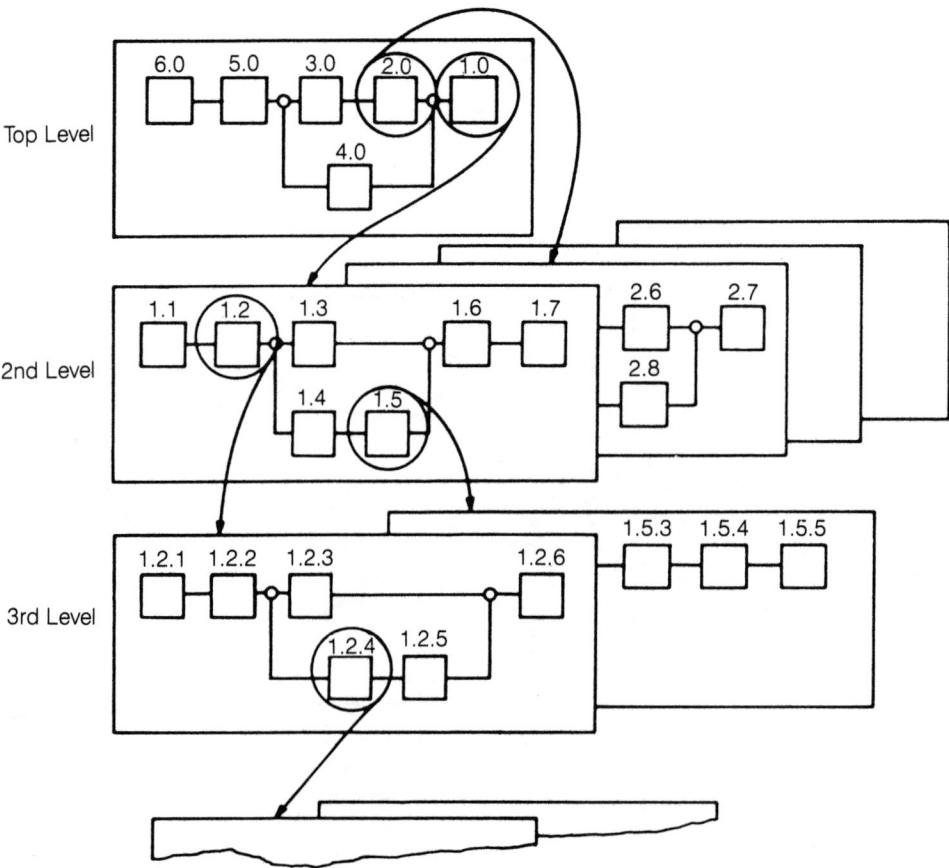

Fig. 2-13. Indenture levels and traceability. (Source: FM 770-78)

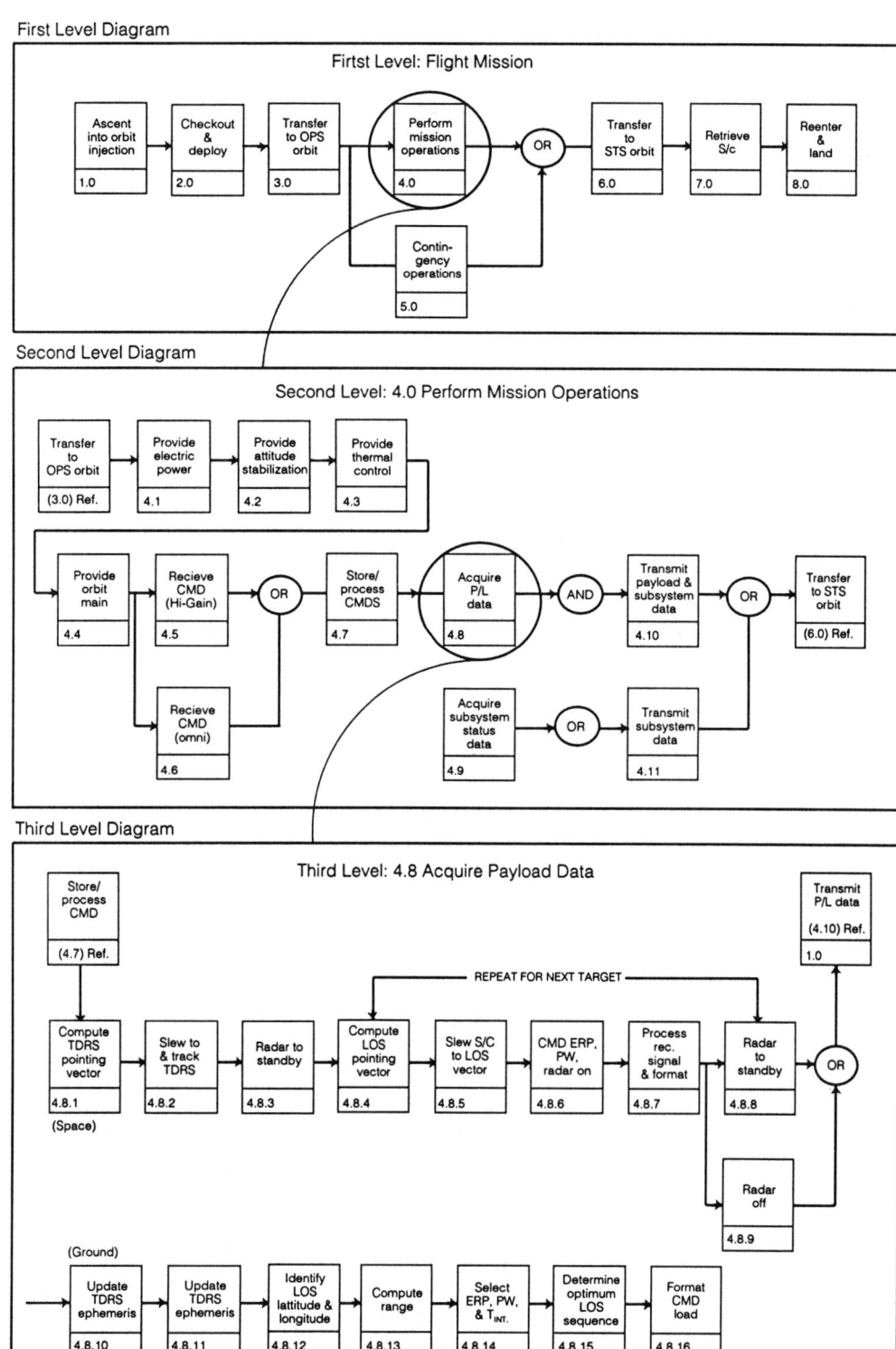

Fig. 2-14. The development of functional flow block diagrams. (Source: Systems Engineering Management Guide, Defense Systems Management College, 1986)

30 The systems engineering process

Time-line analysis supports developing requirements for the product operation, test, and maintenance. The analysis shows:

- Time-critical paths.
- Sequences.
- Overlaps.
- Concurrent functions.

Time-critical functions affect reaction time, downtime, or availability. Performance parameters can be derived, in part, from time-critical functions. Figure 2-15 is an example of a time-line sheet for a maintenance function.

Time-Line sheet	(A) Function — Perform periodic maint on VC distiller		(B) Location — Engine Room 3	(C) Type of maint — scheduled 200 HR PM
(D) Source— FFBD 37.5×3	(E) Function & Tasks— RAS 37.5×37		(F) Time — Hours .5 1.0	
Task SEQ. #	Task	Crew Member		
.01	Inspect compressor belt	A2	.3H	
.02	Lubricate blowdown pump	B1	.2H	
.03	Check mounting bolts	B1	.1H	
.04	Clean breather cap	B1	.1H	
.05	Clean food strainer	C1	.5H	
.06	Replace oil	B1	.2H	
.07	Replace filter	C1	.4H	
.08	Replace V-drive belt	D1	.9H	
.09	Clean & inspect control panel	C1	.1H	
.10	Install new diaphragms	A2	.7H	
.11	Clean controls	B1	.1H	
			Total man-hours—3.6 MH Elapsed time —1.0 H	

Fig. 2-15. Sample maintenance time-line sheet. (Source: Systems Engineering Management Guide, Defense Systems Management College, 1986)

For simple products, most functions are constant and have a fixed relationship to their physical components. This is not the case in more complex products. Here, functions are variables with peak demands and worst-case interactions. The time-line analysis is valuable in identifying overload conditions. A matrix of function needs versus component capabilities to perform the functions can be constructed. The matrix is best left to the analysis activities after the functions have been identified.

Function analysis limits

Unfortunately, function analysis by itself does not adequately describe a product. Function analysis does not describe limitations, iteration, information flow, performance, or environments. However, it is a significant and essential tool in systems engineering activities.

Systems synthesis

Synthesis comes from the Greek "to place with." Synthesis is the opposite of analysis. The composition, combination, and configuration of components concerns synthesis, which collects, invents, and innovates alternatives. It has a strong link to the creative dimension of systems engineering. In this book, synthesis produces alternatives but does not choose the final solution. Sources for collection of alternatives include:

- Literature
- Patent searches
- Experts
- Analogous products
- Brainstorming

Invention is the development of a totally new idea. Innovation is an improvement or combination of existing ideas. Sometimes ideas from unrelated fields find application in unexpected ways. Systems engineers need a broad outlook for the synthesis activity.

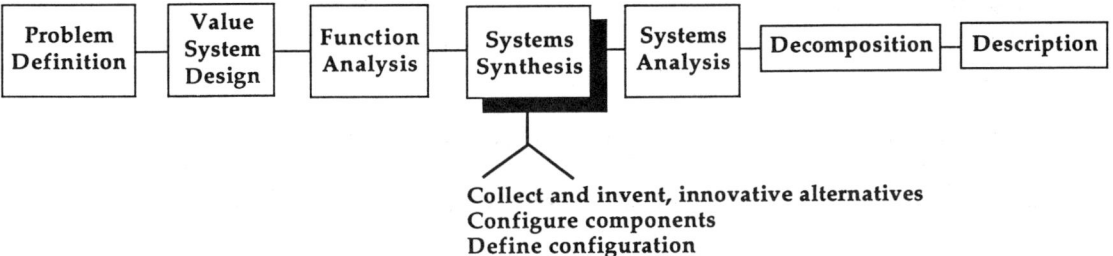

Fig. 2-16. Systems synthesis activity.

Configuring off-the-shelf components into alternatives is synthesis. Systems can be thought of as inputs and outputs linked by functions, timing, and sequences. Within this context, there are many configurations of the same components that might satisfy the customer needs.

Synthesis generates alternatives. There are infinite solutions to the problem. As a practical matter, your own internal cognitive filters will screen out most of the possible solutions. You need to consider more than one solution. Products based on a single-point solution are not competitive in the marketplace unless you are statistically lucky.

Systems analysis

In a sense, problem definition, value system design, function analysis, and systems synthesis are concerned with the "what's." Systems analysis is concerned with the "how."

32 The systems engineering process

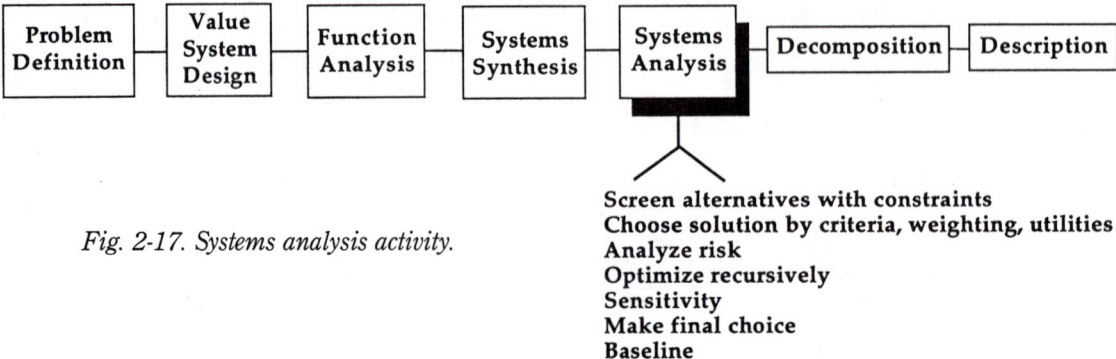

Fig. 2-17. Systems analysis activity.

Decomposition and description, discussed later in the chapter, are concerned with "how much."

Systems Analysis includes choosing a solution from the alternatives, optimizing the solution, and deciding to keep and baseline that solution. This includes:

- Screening alternatives with constraints.
- Choosing by means of criteria, weighting, and utilities.
- Analyzing risk.
- Optimizing recursively.
- Analyzing sensitivity.
- Making the decision.
- Baselining.

Strictly speaking, analysis means to separate into parts or elements. Systems engineering uses analysis to deal with complexity. You deduce the consequences of selecting one of the alternatives by systems analysis. Because you have limited time, the value system is your guide to the priority analyses that must be accomplished. Analysis is facilitated by:

- Defining the problem correctly.
- Selecting the right tool for analysis.
- Knowing the limits of the tool.
- Knowing when to stop.

The right tool is the least costly technique for the needed accuracy. Feasibility studies needing only an order of magnitude do not require high accuracy. On the other hand, trade studies for critical parameters might require very accurate results. When you are selecting a tool, consider the:

- Need for accuracy.
- Time available to produce an answer.
- Cost of the analysis.

Choosing a solution

The three basic steps for choosing a solution are:

1. Screening the alternatives with the constraints.
2. Choosing by using the criteria, weighting, and utilities.
3. Analyzing the risk in using the selected solution.

Because there are infinite alternatives, they must be reduced to a manageable number. The constraints act as a screen to sift out unacceptable alternatives.

Choosing a likely solution

The alternates that pass the test of a likely solution are analyzed using the criteria of the value system. Figure 2-9 shows a simple scoring table. The alternatives are scored against the criteria; the criteria are weighted as to importance; and the alternative with the highest weighted score is considered seriously as a solution.

Scoring can use many different methods. Scores can be as simple as a one, two, three rating. Some scoring uses 0 to 100 as a scale. TABLE 2-2 shows a 0 to 10 rating example.

Table 2-2. Scoring example.

Numerical Score	Definition
10	Exceptional—Exceeds specified performance in
9	a beneficial way, high probability of success,
8	no significant weaknesses
7	
6	Acceptable—Meets standards, good probability
5	of success, weaknesses can be readily corrected
4	
3	
2	Marginal—Fails to meet standards, low
1	probability of success, significant deficiencies but correctable
0	Unacceptable—Fails to meet minimum requirements, needs a major change to make it correct

Source: AF Regulation 70-15

Evaluating alternatives requires a common means of measure. You must compare on a basis of equivalent standards. In addition to the weighted scoring method, you can evaluate alternatives by the Pugh controlled convergence method.

Pugh's controlled convergence

Stuart Pugh of Great Britain developed a technique of selecting the best alternative—controlled convergence. In a sense, you are describing a benchmark and then improving on it. In the process of evaluating alternatives, you also generate new ones.

Pugh's controlled convergence method involves team effort. Pugh's experience is that the method makes it difficult for strong-willed people to push their own ideas for irrational reasons. The peer process is both analytic and synthetic in that both selection and creativity happen. Pugh believes that a disciplined approach leads to improvements in the product development.

The process is recursive, going through several phases to improve the initial concepts. A synopsis of the steps are:

1. Sketching each of the alternatives to the same level of detail.
2. Making a concept evaluation and comparison matrix. An example is shown in FIG. 2-18. Put the sketches on the matrix.

Criteria \ Concept	Sketches					
	1	2	3	4	5	6
A	+	−	+			
B	+	−	−			
C	+	−	+			
D	S	S	S			
Total +	3	0	2			
Total −	0	3	1			

Fig. 2-18. Pugh evaluation matrix example.

3. Choosing the criteria for the selection evaluation.
4. Choosing a benchmark from the alternatives.
5. Comparing the alternatives to the benchmark, sticking to one criterion at a time. Evaluate by the criteria:

   + decidedly better
   − decidedly worse
   S about the same

6. Abstaining from modifying alternatives during the comparison.

7. Adding the pluses and the minuses for each alternative.
8. Looking at the negatives of the strong alternatives. Can they be changed into pluses? Do not change the existing alternatives on the matrix, but add those modified as new additions to the matrix.
9. Looking at the weak alternatives. Can they be saved? If not, delete them from the matrix.
10. Looking for direction of the best alternative, not worrying about numerical scores.
11. Repeating the steps until the design converges to an acceptable solution.

The benefits of this process are:

- New alternatives are created.
- Knowledge of the reasons for concept strength increases.
- The design problems are better understood.

Analyzing risk

Systems engineering evaluates the risk, or potential loss, of selecting an alternative as a solution. Even if a solution is technically the best, social or political considerations might cause the alternative to be a potential disaster. If the potential drawbacks cannot be accepted, the alternative must be discarded or modified.

Chapter 16, Technical Control, covers risk management techniques. Risk analysis is not confined to only the front of the project, but it is a continuing effort.

Optimize Recursively

As you have seen in the Pugh convergence, you can find opportunities to optimize recursively in many activities. The objective for optimization should be to minimize the customer's loss. Usually, in product design, most engineers strive to minimize the manufacturing losses. It's worth repeating: Minimize the customer's loss. When optimizing, you should consider:

- Models and simulations.
- Mathematics.
- Statistics.
- Engineering computations.

Figure 2-19 shows a flow diagram for a general optimization process. The process shown is recursive. Knowledge and information increase with time.

One point you might consider is robust versus optimal. Optimal is not always the best solution. For example, FIG. 2-20 has a function with points "A" and "B" to consider. Point "B" is optimal in that it maximizes y. Point "A" is less in the property of y, but it is more robust. Small variations in x will cause a miss from "B" but have no, or small effect, on point "A." From an engineering and producibility viewpoint, design to point "A" is desirable because it is less sensitive to parameters outside of control.

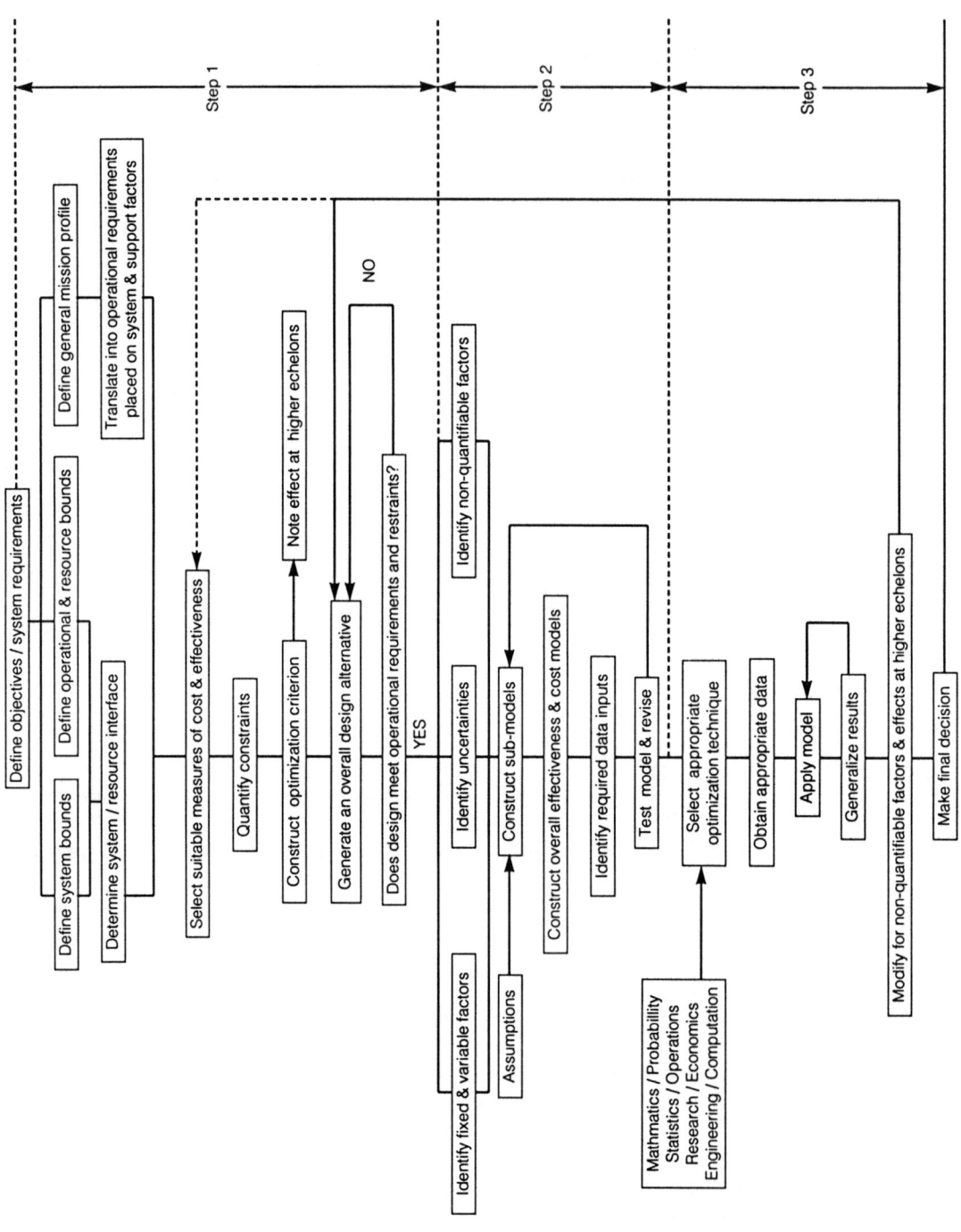

Fig. 2-19. Flow diagram for a general optimization process. (Adapted from: MIL-HDBK-338)

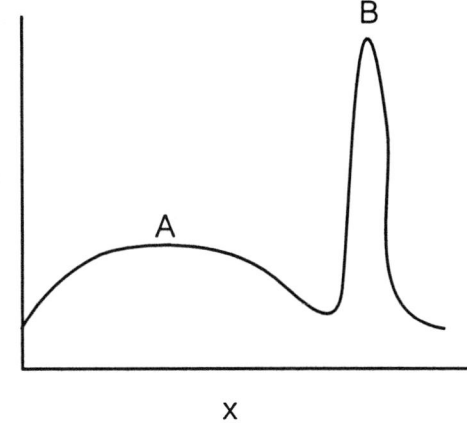

Fig. 2-20. Considering robust and optimal.

Modeling and simulation

Models and simulations allow you to study the effects of choices without actually building and testing a product. A model is a representation of a process or product that shows the effects of significant factors. Simulation uses models to explore the results of different inputs and environmental conditions. Models can be as simple as a picture or sketch. They can also be mathematical and statistical. Beginning models are simple and become more complex as time and understanding increase.

Models are first done by identifying inputs that can be manipulated and what outputs result for the process or product under study. You then examine the effects of the environment on the product's performance. Last, the internal transfer function of the product or process to complete the model is represented. When these are tied together, your model is ready.

Traditional optimization theory uses differential calculus, the simplex method, and other mathematical techniques. Computing power is readily available through desktop computers and spreadsheets. Spreadsheets have built-in numerical functions and iteration capabilities, making them ideal for small models. The references in the Further Reading section are good starting points.

Using statistics

Experimentation time can be reduced by using statistical methods. The term factor is used to denote any feature of the experiment that can be varied, such as time, temperature, or pressure. The levels of a factor are the actual values used in the experiment. Experiments can be designed for best capture of data and reduced number of experiments required. A fractional factorial experiment is an example.

Credit for statistical experiment design goes to R.A. Fisher in England. About 1920, Fisher went to work at an agricultural research station. The researchers were interested in the best ways to grow wheat, potatoes, and other crops. The experiments could have been conducted in a greenhouse under controlled conditions but the results would have reflected a greenhouse and not a farmer's field.

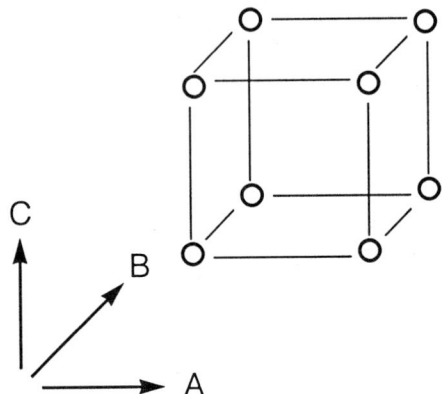

Fig. 2-21. Three-factor, two-level experiment.

Fisher's problem was how to run experiments in a "noisy" and uncontrolled environment. Some fields received rain and others didn't, birds attacked one field and not another, some fields had better drainage than others. One of the ways Fisher solved the problem was to vary all the factors simultaneously in what he called a factorial design. He also used blocking to eliminate the effects of nonhomogeneity, and randomization to avoid confounding with unknown factors.

Fisher's blocking cancelled out the effects of day-to-day variations. The experiments were set up balanced to cancel out variation effects. Unknown trends were taken care of by assigning individual treatment combinations in random order inside the blocks.

Most American engineers have been taught to vary one factor at a time in a simulation, holding everything else constant. This allows understanding of each factor's contribution. Unfortunately, this method takes a great deal of time and does not directly show interactions.

To illustrate the time savings of experimental design, consider a simulation with three factors. You will look at two levels for each factor, hoping to find the best solution. The three-dimensional, orthogonal representation is shown in FIG. 2-21. The factors are A, B, and C. The old way is to run the simulation at every point for a total of eight simulation runs.

Suppose you designed the experiment for only four runs. If the samples were balanced, and you could extract the information statistically, you could save time if you had many factors. Figure 2-22 shows the projection of four samples on three planes. Each of the planes contains the necessary information to extract your desired data. Instead of eight runs, you only need to do four. You only need to learn the statistical technique of setting up the experiment (use tables) and extracting the data (you can use a computer program or do it by hand). There are three advantages of designed experiments:

1. It takes less time to run the simulation or the experiments.
2. Unknown biases are avoided.
3. Variation from day-to-day and batch-to-batch are balanced out.

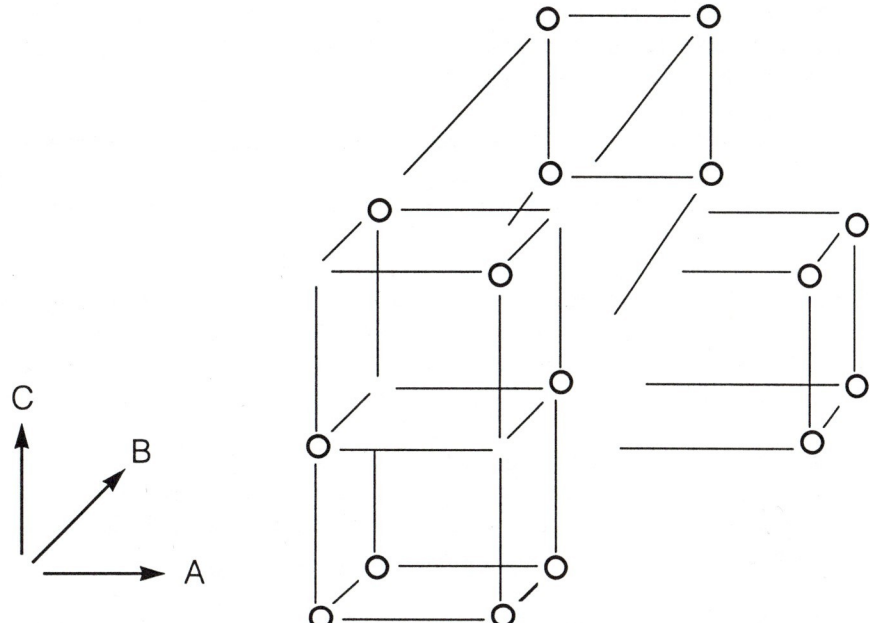

Fig. 2-22. Projection of four samples.

The statistical techniques are not difficult. For engineering work, you can use a cookbook approach to performing the necessary mathematics. Consider asking an experienced person in experiment design for help so that you measure the factors properly.

Using Taguchi methods

Japanese engineers have been using Dr. Genichi Taguchi's quality engineering methodology for more than 30 years. Commonly known as Taguchi methods in America, the methodology uses two primary tools, the Signal-to-Noise Ratio and the Quality Loss Function. The idea is to develop high-quality, low-cost products that are insensitive to variability-causing factors in manufacturing and in the field. This is a change from the American thinking of pass/fail to specification limits.

Dr. Taguchi believes that low cost should be a precondition in raising quality levels. Price competition is primary. You might have perfect quality, but if the price is high, then the competition will beat you. He believes that the right way to produce products is to get the cost down, improve quality as much as possible by parameter design, and then perform tolerance design, if necessary, at the end. Therefore, the cost of quality must be calculated.

The designer has to find solutions to quality and cost problems caused by many factors, including those he knows nothing about. The Taguchi method uses statistics to analyze the main parameters and how to make use of their interactions with unknown causes. Mathematicians find some fault with Taguchi methods because they are not mathematically rigorous. Taguchi's reply is that engineering is a completely different world from science in that engineering problem-solving uses short cuts to get practical answers, not perfect answers.

It is interesting that Taguchi methods do not seek to identify cause-and-effect relationships. Understanding causes is not necessary to design robustly against variation-causing influences. The methods do place a strong reliance on the product knowledge of the engineer.

The Taguchi method use an analogy from communications engineering. That is, the Signal-to-Noise ratio. Product characteristics should have good signal-to-noise ratios. That is, they should be impervious to noise. Noise factors are anything that the engineer can't control. Noise causes quality characteristics to deviate from the target, which causes a loss. Three types of variability-causing noise are:

1. External noise—variables in the environment or conditions of use.
2. Internal noise—changes that occur when a product deteriorates or ages.
3. Unit-to-unit noise—differences between individual units that are manufactured to the same specifications (manufacturing noise).

The engineer does not attempt to control the noise factors. This is usually expensive. The engineer designs around the noise factors, choosing parameters that minimize the effects of the noise.

Taguchi methods consider three stages for both the engineering of the product and the manufacturing system to build the product:

1. System design—the functional design.
2. Parameter design—the most important work after the system has been decided—minimize product variation to noise—design that minimizes the need for tolerances.
3. Tolerance design—might improve function and quality, but adds to the cost.

The manufacturing system design to satisfy the specifications of the product include the:

- System design—the manufacturing processes selected from technology.
- Parameter design—the optimum working conditions, including materials and purchase parts.
- Tolerance design—the tolerances of the process conditions and sources of variability that are set.

The Quality Loss Function describes the loss to the customer for deviation from the target values. American specifications call for a pass/fail test for conformance. Taguchi shows that any deviation from target is a loss to the customer. Taguchi uses the example of a student. One student makes a grade of 60 and passes, another a grade of 59 and fails. Is there a step function in knowledge between them? No. There is a curve of loss from the target value of 100.

Taguchi uses a quadratic loss-curve equation to compute the loss to the customer. The on-target loss is zero. The costs as the product moves away from target are based on a tangible cost such as warranty costs. The curve can be fitted to pass through such tangible cost points. The objective of Taguchi methods is to minimize the loss to the customer.

Engineering minimizes losses by selecting a low-cost system design. Key parameters must be identified that allow the least variation in the presence of noise. This is done by using experiments, usually orthogonal arrays. The levels of the parameters are set for least variation, again using orthogonal arrays. The results are confirmed before engineering release. Concentration is on the "vital few" parameters. Only those parameters that can be controlled in a cost-effective manner are used.

The Taguchi method is a serious means of looking at engineering. The results of the Taguchi methods have also been proven in the marketplace and are a tool for cost reduction and customer satisfaction.

Analyzing sensitivity

Analyzing sensitivity means analyzing the sensitivity of the proposed solution to changes in the value system, requirements, or functions, as well as identifying changes in weights or scoring that might reverse decisions. Utility curves often point out peaks of optimization that might not be stable, and analyzing sensitivity can prevent selecting an unstable design.

You might want to use optimization methods and designed experiments to determine sensitivities to changing environments and other noise. Manufacturing methods are another area you might want to cover.

Making the decision

You have designed a value system and used it to judge the alternatives. You have optimized and evaluated for risk. You now must make your decision based on a systems engineering logic and not from a guessed, single-point solution.

Trade studies

Trade studies document the systems analysis activities. The same activities you have looked at in this chapter are used to perform trade studies. Always define your objectives first before analyzing alternatives.

Large analyses can be broken down using trade trees. The trade tree uncouples many alternatives into several smaller studies. This is a good approach to use when starting at the top of the product hierarchy rather than evaluating every possible alternative at the bottom of the alternative tree. When evaluating a trade study, ask yourself these questions:

- What requirement or function does this study?
- What are the confidence levels of the information used?
- Are all reasonable alternatives being explored?
- Is a matrix of alternatives and criteria presented?

- Are all significant criteria identified?
- What is the rationale for the weighting?
- Do utility curves exist?
- Has risk been considered?
- What are the sensitivities?

A written trade study report might include the:

- Purpose of study
- Scope of report
- Background information
- Alternatives
- Constraints
- Criteria and weighting
- Evaluation method
- Risk
- Conclusion/decision
- Information sources

Systems engineering maintains a trade study report index, which allows rapid retrieval of decision documents. Number your project's trade studies for cross-indexing into the requirements allocation sheets.

Baselining

Baselining your decisions means taking a snapshot of the design at a given point in time. Engineering is recursive and changes will happen. Having a baseline makes it easier to understand the effects of new changes, both in requirements and design.

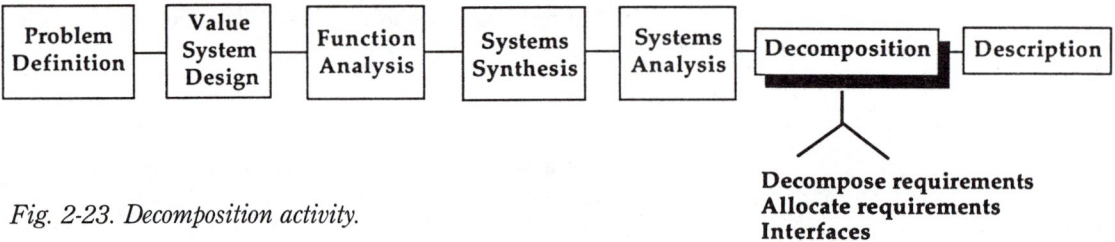

Fig. 2-23. Decomposition activity.

Decomposition

Decomposition activities include decomposing requirements, allocating requirements, and defining interfaces.

Decomposing and allocating requirements

Decomposing requirements means that you break down requirements past the components, even to the manufacturing process. Allocating requirements is assigning portions of a requirement among ensuing lower levels of the product hierarchy. Derived requirements are those that change from higher-level requirements to a parameter that is indirect. For example, a customer requirement for "easily seen" might translate into a requirement for orange markings.

Requirements derivation might be needed when you cross hierarchal boundaries. For example, a requirement for electromagnetic compatibility might specify a field at the system level. Within the metallic box of that system, an electronic board might see less field strength because of the shielding of the box. Therefore, the levels inside the box will have to be calculated or measured.

Another example of derivation is the mission profile. The mission profile has two parts: (1) functions that must be performed shown on a time scale; and (2) environmental properties and limits shown on a time scale. You might consider doing these profiles not only at the system level but also at each hierarchal level as the functions are broken down and the environment changes because of enclosures.

Decomposition examines the role of the product and the role of the people using the product. Determining the roles and interactions of people and the product are important decisions. You also break down tasks performed by people. Human engineering specialists derive subtasks from primary human tasks performed with the product. Task analysis gives a detailed sequential description of: time-shared tracking or continuous tasks and procedural or discrete tasks.

Quality Function Deployment

Quality Function Deployment (QFD) is a useful tool for decomposing requirements. It is also an excellent tool for planning requirements. Quality Function Deployment integrates many of the systems engineering activities and tools. Interestingly, Quality Function Deployment began in Japan about the same time that J. Douglas Hill and John Warfield published a paper called "Unified Program Planning" in 1972 that describes linking correlation and self-correlation matrices. QFD might be based on systems engineering, but it integrates the planning and flowdown beautifully. QFD provides information including answers to:

- What is important to the customer?
- How is it provided?
- What relationships are there between the "whats" and "hows?"
- How much must be provided by the "hows" to satisfy the customer?

The first step is determining the "whats." At the top-product level, the whats are taken directly from the customer. Information such as "must work a long time without breaking" is organized into categories. This demanded quality is organized into 1st level, 2nd level, and 3rd level with the lower levels supporting the next higher level. An importance rating is assigned to each demanded quality. Prioritizing is the most important activity in Quality Function Deployment.

With the whats organized and rated, the next step is to describe the hows. The hows are allocated and derived requirements. At the top-product level, the hows describe the product features and characteristics. The whats and hows are linked by a correlation or relationship matrix. There is no one-for-one relationship between the whats and hows. The matrix allows you to see unfulfilled customer demands and also features that are expensive yet do not serve the customer.

The hows have their own self-correlation matrix. This shows requirements that might reinforce or oppose each other. The hows are given target values for "how much." Engineering does a competitive assessment on the "how much" against benchmarks.

At the next lower hierarchal level, the whats come from the higher-level hows. The requirements flow down in this manner. Quality is deployed from the voice of the customer through marketing, engineering, manufacturing, and supporting organizations.

Figure 2-24 illustrates the organization of a Quality Function Deployment chart. Charts should be kept small, about 30 × 30. Use the Pareto rule. Don't ask customers about things they don't know. An incomplete chart will do more harm than good for you.

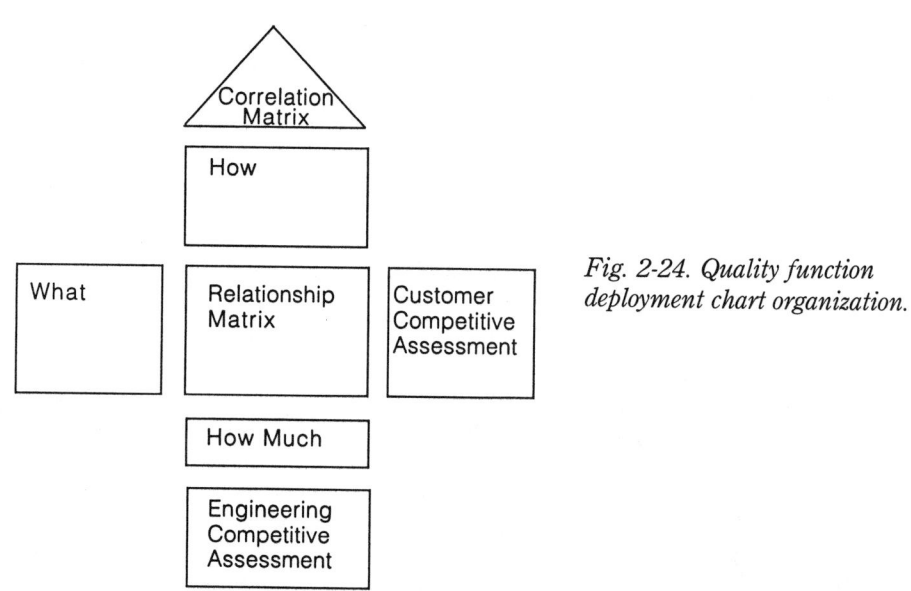

Fig. 2-24. Quality function deployment chart organization.

You can purchase software that organizes and prints the charts. There is a standard symbology for relationships and correlations. If you are a first-time user, your goal might be to just get through the first time. Mastery of the chart technique takes practice.

Defining interfaces

As you decompose requirements and functions, interfaces appear. These are simply boundaries between areas. Defining interfaces allows you independence of development. Clearly defined interfaces reduce risk.

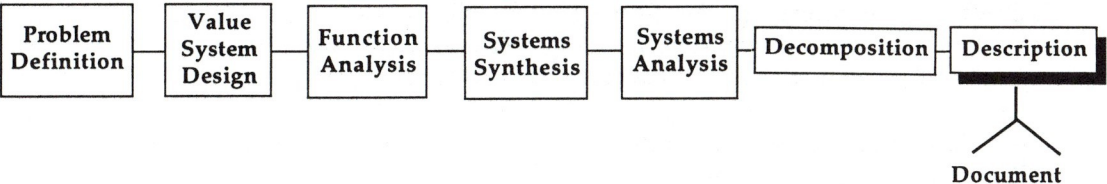

Fig. 2-25. Description activity.

Description

The description activity is the documentation of your efforts. It includes specifications, design sheets, traceability, and configuration control. Specifications, design sheets, and configuration control are covered in chapter 16, Technical Control.

Quality Function Deployment leaves a document trail for tracing requirements. Another method is through the Requirements Allocation Sheet. A Requirements Allocation Sheet (RAS) defines each requirement, identifies its source, and shows the allocation to the next lower level. One means to accomplish this is to use a relational database. A typical RAS may contain:

- Requirement identifier.
- Requirement description.
- Source of the requirement.
- Parent or next upper-level requirement.
- Allocation or next lower-level requirement.
- Verification method and hierarchy level.
- Specification number and paragraph number for this requirement.
- Work breakdown structure reference number.
- Function number.
- Trade study number.
- Mission profile number.

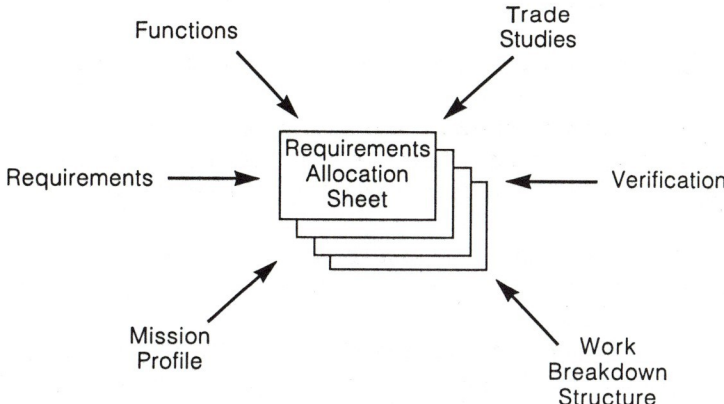

Fig. 2-26. Requirements allocation sheet.

The reason these records are kept is to have traceability of requirements should any requirement change. Requirements Allocation Sheets do not derive or allocate requirements, they are strictly a documentation system for verifying that:

- All requirements have been allocated.
- Requirements will be verified.
- Requirements can be traced up and down the hierarchal levels.

Handling changes

Your system engineering *system* should be robust and easily capable of handling change. Change will happen, because engineering is a recursive process. When designing your system to design the product, ask yourself:

- When does change occur?
- Where does it happen?
- How should changes be handled?

Changes occur as knowledge and information increase. The customer and the market might change. New technology might take over the old. When a constraint is found as you go down the hierarchy, the need for change might become apparent.

Customer and market changes can cause a change at the top of the hierarchy, forcing change to flow down through every level. Therefore, traceability of requirements is needed from the top down.

When a constraint is found at a lower level, you must be able to trace the source of the requirement from the bottom up. You need traceability of your trade studies to backtrack to adjacent alternatives, or close to the solution space of the originally chosen alternative.

Following the system engineering activities allows you to handle change. You have documented requirements, values, functions, alternatives, and means of decisions. You have these at each level of the product hierarchy so that you can make changes to the lowest possible level. A suggested sequence for change is:

- Reverify design adequacy through analysis or empirical proof.
- Redesign to correct problems.
- Reallocate design requirements to correct allocation errors.
- Redefine design requirements to move to an adjacent solution space.
- Reevaluate customer requirements to fall within constraints.

These steps should be worked from the lowest level of the product hierarchy affected by change upwards until the issue is resolved. This causes the least change in most cases.

Make sure the system communicates changes to the rest of the project team. Changes affect not only the product but also the support system, manufacturing, maintainability, documentation, test, and possibly interfaces.

Changes in the work to be done require changes in the work packages. It is important to define what is included and what is not included in work packages for later redefinition of the work effort because of change.

Summary

- Four dimensions of systems engineering are covered. They are:
  - Logic.
  - Information.
  - Creativity, innovation, and invention.
  - Time.
- The general logic dimension of systems engineering includes:
  - Problem definition.
  - Value system design.
  - Function analysis.
  - Systems synthesis.
  - System analysis.
  - Decomposition.
  - Description.
- Some of the sub-activities of the Problem Definition task are:
  - Needs assessment.
  - Developing requirements.
  - Constraints.
  - Scenarios.
  - Scope.
  - Identifying key people needed.
  - Identifying the degree of need.
  - Identifying alterables.
  - Bounding and partitioning the problem.
  - Environments.
  - Describing interactions.
  - Determining normal and unnormal conditions.
- The customer will tell you many of the requirements. Typical ways of listening to the customer are through:
  - The customer interview.
  - The customer questionnaire.
  - Consumer panels.
  - Product testing services.
- Joint Application Design (JAD) is a common effort used by system users and system designers. It centers about a structured workshop called the JAD session.
- The N × N chart shows interfaces and relationships. It is usually used for data, process, or hardware interfaces.
- Quality Function Deployment (QFD) is an excellent tool for both planning and requirements flowdown. It combines elements of the cross-correlation chart and the self-interaction matrix.

- The value system has objectives, objective measures, criteria and weighting, and utilities.
- Constraints are the "musts," the restrictions, the limitations that must be met.
- Criteria are continuous. Criteria are a way of judging the feasible alternatives. Weighting factors are assigned to criterion indicating importance.
- Utility describes the relative value of a criterion for different levels of performance.
- Function analysis has two extremely important benefits. It discourages single-point solutions. It describes the desired behaviors that become requirements.
- Functions are expressed in two words—an active verb and a noun. Functions are expressed as what the customer desires.
- Supporting functions ensure reliability, make the product easy to use, delight the customer, or please the senses, for example.
- The advantages of designed experiments are:
  ~ Less time is required to run the simulation or the experiments.
  ~ Unknown biases are avoided.
  ~ Variation from day-to-day and batch-to-batch are balanced out.
- Japanese engineers have been using Dr. Genichi Taguchi's quality engineering methodology for more than 30 years. Commonly known as Taguchi methods in America, the methodology uses two primary tools, the Signal-to-Noise Ratio and the Quality Loss Function.
- The mission profile has two parts: 1) functions that must be performed shown on a time scale; and 2) environmental properties and limits shown on a time scale.
- A Requirements Allocation Sheet (RAS) defines each requirement, identifies its source, and shows the allocation to the next lower level.

Further reading
Systems engineering methodology

Athey, Thomas H. *Systematic Systems Approach: An Integrated Method for Solving Systems Problems.* Englewood Cliffs, NJ: Prentice-Hall, 1982.

Chase, Wilton P. *Management of System Engineering.* New York: John Wiley & Sons, 1974.

Chestnut, Harold. *Systems Engineering Methods.* New York: John Wiley & Sons, 1967.

Hall, Arthur D. *A Methodology for Systems Engineering.* Princeton, NJ: D. Van Nostrand, 1962.

Hall, Arthur D. III. "Three-Dimensional Morphology of Systems Engineering." *IEEE Transactions on Systems Science and Cybernetics,* Vol. SSC-5, April 1969, pp. 156-160.

Hill, J. Douglas and John N. Warfield. "Unified Program Planning." *IEEE Transactions on Systems, Man, and Cybernetics,* Vol. SMC-2, November 1972, pp. 610-621.

Sage, Andrew P. *Methodology for Large Scale Systems.* New York: McGraw-Hill, 1977.

Mission requirements

NASA. *Introduction to the Derivation of Mission Requirements Profiles For System Elements.* NASA SP-6503. Washington, D.C.: NASA, 1967.

Joint application design

Wood, Jane and Denise Silver. *Joint Application Design: How to Design Quality Systems in 40% Less Time.* New York: John Wiley & Sons, 1989.

Pugh controlled convergence

Pugh, Stuart. *Total Design: Integrated Methods for Successful Product Engineering.* Wokingham, England: Addison-Wesley, 1990.

N × N chart

Lano, R.J. *A Technique for Software and Systems Design.* Amsterdam, The Netherlands: North-Holland Publishing Co., 1979.

Optimization

Orvis, William J. *1-2-3 for Scientists and Engineers.* Alameda, CA: SYBEX, 1987.

Pierre, Donald A. *Optimization Theory with Applications.* New York: Dover Publications, 1986.

Taguchi methods

American Supplier Institute. *Taguchi Methods: Selected Papers on Methodology and Applications.* Dearborn, MI: ASI Press, 1988.

Phadke, Madhav Shridhar. *Quality Engineering Using Robust Design.* Englewood Cliffs, NJ: Prentice Hall, 1989.

Taguchi, Genichi. *Introduction to Quality Engineering: Designing Quality into Products and Processes.* Tokyo, Japan: Asian Productivity Organization, 1986.

Quality function deployment

Akao, Yoji, ed. *Quality Function Deployment: Integrating Customer Requirements into Product Design.* Cambridge, MA: Productivity Press, 1990.

3

Verification

System engineering verifies that the product meets the customer's needs and expectations. Customer satisfaction means continued business and funds for more product development. In this capacity, the system engineer acts as the user and customer advocate. In this chapter, you'll learn about:

- What to verify.
- Means of verification.
- Types of testing.
- Reducing integration and test time.

Defining verification

Verification asks the question, "Are we building the product right?" Validation asks "Did we build the right product?" System engineering ensures that verification happens. The customer validates the product. System engineering verifies both the product and the process of engineering the product. You can verify the process by looking at the logic activities. This is done by holding design reviews and walkthroughs at critical design milestones. Models and simulations are used to prove alternatives. Plans must be made and resources available to complete tasks. Verification is more than checking parts, however. It also:

- Reduces risk early in the project.
- Provides feedback to designers before product delivery.
- Proves working interfaces.
- Examines behaviors.
- Confirms functions.
- Proves requirements are met.

Problems do and will happen in engineering. Verification provides information about faults so they can be fixed and catches problems early while the cause can still be traced. Problems caught later usually mean something must be expedited to regain schedule. This adds cost to the product.

Because problems happen, use the principle of independence. Keep activities and components independent of each other as much as possible. Precedence restrictions and co-dependence will add to project schedule time.

Early verification saves money. The cost of finding a fault rises about 10 times for each level of hierarchy. For example, finding a faulty diode at incoming inspection might cost less than a dollar. The same fault at a system level could cost $1,500 for complex products. Costs rise with complexity and increased troubleshooting, teardown, reassembly, and retest. Therefore, find problems as early as you can.

Consider verification when specifying requirements. One of the attributes of a good requirement is the ease of verification. Plan verification by means of an integrated test plan.

Verifying the product

Testing increases confidence in meeting customer requirements. It is part of risk reduction. Verifying performance through testing means detecting:

- Designs and design changes that fail to meet requirements.
- Manufacturing defects.
- Component failure or nonconformance.

Figure 3-1 shows the progression of product development. The product is defined first at the highest hierarchal level and then requirements flow down to the lower levels.

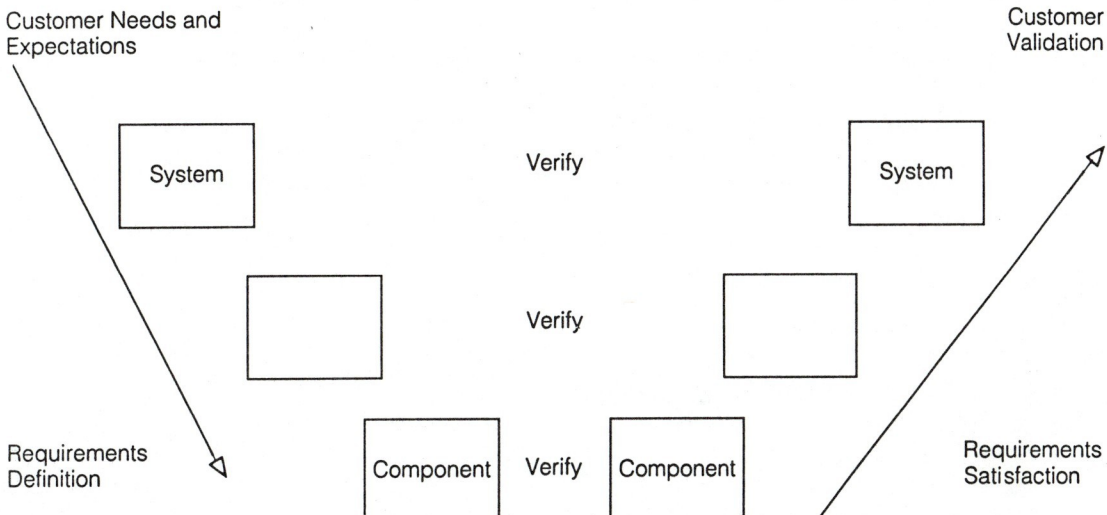

Fig. 3-1. Verification at all levels.

52 The systems engineering process

The customer requirements are transformed into engineering requirements to manufacturing. Manufacturing begins making materials and small parts. The parts are assembled into components of increasing complexity until the final product is complete. Verification occurs throughout this approach.

Planning and controlling

Planning for verification includes:

- Design and acquisition of test equipment.
- Determination of floor space, facilities, and people.
- Determination of schedule.
- Basis for detailed test procedures.
- Integration of verification activities.

Verification controls should ensure that:

- Verification documents are properly supervised and approved.
- Verification documents are under configuration control.
- Nonconformance is identified and analyzed.
- Measuring and test equipment is calibrated to a traceable standard.

Verification methods

Verification methods usually fall into one of five categories:

1. Inspection—Visual, sound, smell, taste, or touch examination to determine compliance with the requirements. Might use gauges or simple measurements.
2. Analysis—Technical evaluation of data, logic, or mathematics to determine compliance with requirements.
3. Demonstration—An uninstrumented test where compliance is determined by observation. For example, time to perform maintenance tasks.
4. Test—Using procedures and equipment to verify compliance with requirements.
5. Process control—Accepting process control as evidence that requirements are met. Process factors are understood, measured, and held to predetermined target values.

Testing

Most types of tests fall under:

- Burn-in or environmental stress screening.
- Environmental.
- Variable versus go-no go.
- Hierarchal level.
- Production assessment.
- Destructive versus nondestructive.

Verification

Burn-in is testing during temperature cycling and vibration. Temperature cycling stresses the product. Random vibration causes loose screws and parts to work free. Burn-in screens finds nonconformance due to:

- Parts failure.
- Manufacturing defects.
- Marginal design.

Environmental testing verifies performance in the customer's environment. Environments include:

- Altitude
- Temperature
- Solar radiation
- Rain
- Humidity
- Fungus
- Salt fog
- Sand and dust
- Explosive atmosphere
- Water immersion
- Acceleration
- Vibration
- Acoustic noise
- Shock
- Icing and freezing rain
- Electromagnetics

Variables testing records the actual value of the measurement. Go-no go compares the measured value against predetermined limits and decides if it is acceptable or not.

Hierarchal level refers to the level of assembly where testing is performed. You already know to test and catch problems at the earliest opportunity. Some problems that might not show until higher assembly include tolerance build-up, race conditions, sneak paths, and stored energy hazards. For example, paralleling relays without isolation diodes will cause "chattering" relays because of stored charge in the relay coils.

Production assessment testing is done on samples drawn periodically from production of the product. This is an on-going verification. Examples are expensive tests such as environmental testing. Another example is verification of weight when the product is under configuration control. It is a check on processes and parts that might change undetected.

Destructive tests render the test object unfit for its intended use. These tests must be done as samples, or nothing would be left to ship. Destructive tests are done on objects such as fuses, flash bulbs, and metallic materials.

Test and evaluation

Verification ensures that a product meets the requirements. It would be nice to have confidence that the product will work before it is assembled, however. The purpose of "test and evaluation" is to identify areas of risk for elimination or reduction during the product's development. You could say that test and evaluation is more risk reduction than verification but it is a verification of the system engineering process. That is, by reducing risk, you are minimizing loss to society.

Test and evaluation generates information and knowledge on the developing product. It is deliberate and rational. System engineering compares and evaluates results of testing against the requirements. Test and evaluation includes physical testing, modeling and simulations, experiments, and analyses.

"Test" means the actual testing of the product and components. "Evaluation" is the review and analysis of the information. The distilled information allows system engineering to:

- Define requirements.
- Manage the system engineering process.
- Identify risk.
- Discover new alternatives.
- Improve product robustness.
- Find constraints.
- Decide the allocation of resources.

Design for testing

Efficient test and evaluation demands design for testing during product development. System engineering addresses the need to:

- Collect data during the development process.
- Enable easy measurement, including:
  - Partitioning
  - Controllability
  - Observability
- Enable rapid and accurate assessment of the information.

Integrating test and evaluation

Test and evaluation must be integrated with the rest of the system engineering effort. Documented decisions for test and evaluation are called the Test and Evaluation Master Plan (TEMP). The testing program in the TEMP must be consistent with the System Engineering Management Plan (SEMP). The testing program in the TEMP must provide the technical performance measurements required for reviews, audits, and risk management. Other documents integrated with the TEMP include the:

- Configuration management plan.
- Functional analysis documents.

- Requirements Allocation sheets.
- Test Requirements sheets.
- Specifications.

Test and evaluation is not limited to the primary product. The facilities and support system need risk reduction also. For example, supportability can and must be measured.

Reducing integration and test time

You want thorough verification, yet at the same time your schedule is breathing down your neck. How do you reduce integration and test time without reducing quality? Perhaps TABLE 3-1 has the clue. As the requirements flow down to lower levels, a future verification is also created. By considering future verification and integration while defining requirements, you can design the product for reduced cycle time. Ideas for implementation are shown in TABLE 3-1.

Table 3-1. Reducing Integration and Test Time Ideas.

- Easily verifiable requirements.
- Interface definition.
- Peer walk-throughs.
- Models and simulations.
- Robust design to component parameter variation, manufacturing process.
- Robust inputs, target outputs.
- Commonality, standardization.
- Simplicity.
- Testability.
- Reliability.
- Maintainability.
- Test equipment and facilities available.
- Independence of components.
- Hardware emulator for untested software; tested software for untested hardware.
- Modular, bottom-up testing.
- Understand the critical path.
- Have test plan, test procedures ready.

Summary

- Verification asks the question, "Are we building the product right?" Validation asks "Did we build the right product?"
- Early verification saves money. The cost of finding a fault rises about 10 times for each level of hierarchy.

- Verification methods usually fall into one of five categories:
  - Inspection
  - Analysis
  - Demonstration
  - Test
  - Process control

Part II

Specialty engineering

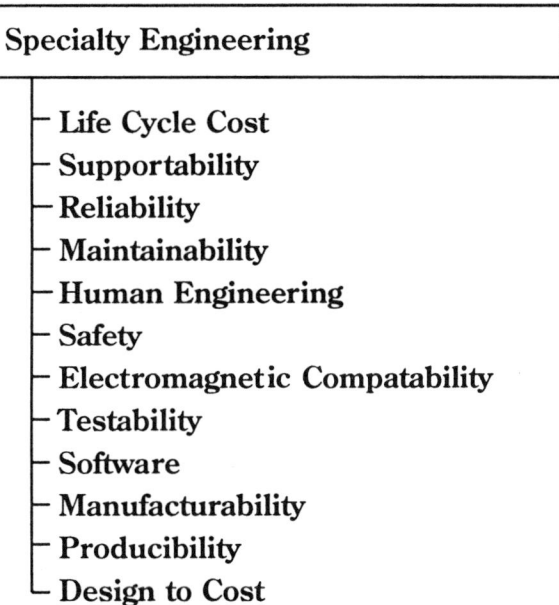

Engineering specialties apply knowledge from specific areas into the logic dimension. They include reliability, maintainability, safety, and others. Specialty engineers draw on a body of knowledge taken from past and current projects. Specialists also help define requirements, audit and verify that requirements are being met, and act as expert resources to designers so that requirements can be met.

Systems engineering integrates designers and specialists toward the common goal of satisfying the customer. It ensures that the value system is correct and that tasks and activities contribute value to the product.

Systems engineering must coordinate and open communications among the engineers. It is a good idea to keep the perspective that you are integrating the efforts and not doing all of the details yourself. In Part II, you'll learn about the tasks done by the specialists. The purpose of Part II is to show why specialty tasks are important and what some of them are. Management of the diverse tasks will be held for Part III.

4

Life cycle cost

Life cycle cost is the total cost of acquiring and utilizing a system over its entire life span. It is also a major factor in market success. Your job, as the system engineer, is to minimize that cost. In this chapter, you'll learn:

- How to drive down cost.
- What tasks the life cycle cost specialist performs.
- How other specialty engineering affects life cycle cost.

Life cycle cost allows decisions for:

- Lowest cost.
- Sensitivities between cost and parameters.
- Trade-offs in the requirements process.

Defining life cycle cost

Life cycle cost is the total cost to the customer of acquisition and ownership of a system over its full life. It includes the cost of development, production, operation, support, and disposal. The life cycle cost specialist works in a less than exact world because the tasks are highly estimative. Life cycle cost tools are useful for comparison and trade-offs, but are not as useful for budgeting. Use the life cycle cost specialist to find minimum system total cost.

Numerous system parameters and factors influence life cycle cost. These include:

- System performance requirements.
- Reliability/maintainability requirements.
- Technology.
- System complexity.
- Production quantities.
- Production learning curves.
- Maintenance and logistic support plans.

Specialty engineering

Life cycle cost organizes cost categories according to the system's life phases. Figure 4-1 shows the relationship of the major categories. Figure 4-2 breaks out the costs to a lower level. TABLE 4-1 shows the cost breakdown for software.

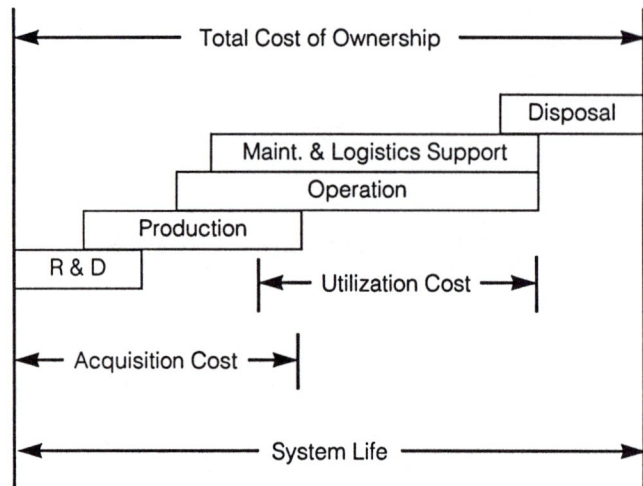

Fig. 4-1. LCC categories vs. life cycle. (Source: MIL-HDBK-338)

The life cycle cost specialist uses Cost Estimating Relationships to estimate the cost of individual elements in the cost breakdown structure. Each Cost Estimating Relationship (CER) contains variables and parameters reflecting conversion factors and empirical relationships. The three basic methods used to develop CERs and to estimate cost are the:

1. Engineering cost method.
2. Analogous cost method.
3. Parametric cost method.

The engineering cost method is a bottoms-up estimation. It uses standard established cost factors from supplier quotations, manufacturing estimates, and best-judgment engineering estimates. It is usually the most detailed estimate.

The analogous cost method bases estimates on past experiences with similar equipment, technology, and software. It uses historical data, which is factored for inflation, and the effect of technology advances.

The parametric cost method uses significant parameters and variables to develop CERs. A parameter is a conversion factor from one system of units to another. It might be a price, an empirically derived ratio, or a policy parameter. An example of a policy parameter is the number of support shops. A variable characterizes resource consumption over time. Variables generate costs. An example of a variable is the failure rate.

Total Life Cycle Cost

Acquisition

Basic Engineering
- Design (electrical, mechanical)
- Reliability, maintainability
- Human factors producibility
- Component
- Software

Test and Evaluation
- Development
- R growth
- R&M demonstration
- R screening
- R acceptance

Experimental Tooling
- System
- R program (MIL-STD-785)
- M program (MIL-STD-470)
- Cost

Manufacturing and Quality Engineering
- Process planning
- Engineering change control
- Q.A. planning, audits, liaison, etc.

Recurring Prouction Costs
- Parts and materials
- Fabrication
- Assembly
- Manufacturing support
- Quality control
- Inspection & test
  Receiving
  Inprocess
  Screening
  Burn-in
  Acceptance
- Material review
- Scrap rate
- Rework

Nonrecurring Production Costs
- First article tests
- Test equipment
- Tooling
- Facilities
- System integration
- Documentation (including maintenance instructions and operating manuals)
- Initial spares (organizational, intermediate, and depot) (pipeline)

Operation & Support

Logistics & Maintenance Support
- Pipeline spares
- Replacement spares (organization, intermediate, depot)
- On-equipment maintenance
- Off-equipment maintenance
- Inventory entry & supply management
- Support equipment (including maintenance
- Personnel training & training equipment
- Technical data & documentation
- Logistics management
- Maintenance facilities & power
- Transportation (of failed items to and from depot)

Operational
- Supply management
- Technical data
- Personnel
- Operational facilities
- Power
- Communications
- Transportation
- Materials (excluding maintenance)
- General management
- Modifications
- Disposal

Fig. 4-2. Life cycle cost breakdown. (Source: MIL-HDBK-338)

Table 4-1. Software Life Cycle Cost Breakdown Structure.

Analysis requirements
    System
    Program
    Interface
    Design
Design
    Flow charts
    Data structure
    Input/output parameters
    Test procedures
Code and checkout
    Coded instructions
    Code walk-throughs
    Compile program
Test and integration
    Program test
    System integration
Documentation
    Listings
    User manual
    Maintenance manual
Installation
    Validation
    Verification
    Certification
Operation and support
    Environments
    Modifications
    Document revisions
    Test revisions

Source: U.S. Department of Defense. *Electronic Reliability Design Handbook.* Military Handbook 338. Washington, D.C.: U.S. Department of Defense.

Equations combine the cost estimating relationships. The cost equations estimate costs by modeling. Life cycle cost models should have certain characteristics including:

- All significant cost drivers.
- Input data that is readily available.
- An allowance for inflation and the learning curve.
- Inputs and outputs in terms familiar to users.
- The results of modeling should be repeatable.

Models cannot make up for poor input data. Data is poorest at the beginning of the project because of uncertainty. Input data will come from the:

- Product description.
- Product life.
- Time phases of the project.

- Projected production quantities.
- Meantime between failures.
- Testability.
- Maintainability and meantime to repair.
- Maintenance concept.
- Hours of operation.
- Personnel required for operation and maintenance.
- Spares and spares provisioning.
- Training.
- Facilities.
- Publications.
- Engineering costs.
- Manufacturing costs.
- Standard cost factors.
- Many other elements.

As you can see from the input data, reliability, maintainability, testability, human engineering, and producibility strongly influence life cycle cost.

Benefiting from life cycle cost

Life cycle cost benefits the systems engineer by providing:

- A basis for evaluating alternative systems and design parameters.
- A method for establishing development and production goals.
- A basis for budgeting.
- A framework for system and project management decisions.

Life cycle cost analysis provides an effective tool to evaluate trade-offs in the design process. It is extremely important that the data inputted be as accurate as possible. Reliability and system support should be analyzed for sensitivities to cost over the life of the product. Challenge requirements from the view of the total system for cost effectiveness. That is, make sure you have the right requirements for design, manufacturing, operation, and support.

Goals may be more appropriately expressed as intervals in the early project stages. Single-point life cycle cost estimates may be less than completely accurate.

Life cycle cost estimates are not good for budgeting directly, but they do provide a basis for what is reasonable. Life cycle cost estimates are better for comparisons than absolutes.

Early project decisions have a heavier impact on the ultimate life cycle cost of a system than those made later. Early life cycle cost information has more leverage and should be used for decision making as soon as possible.

Life cycle cost models pinpoint cost drivers. Use these models to identify hardware, software, and input parameters having the greatest cost impact. Rank these areas by cost for further trade-off and analysis.

Tasking the life cycle cost specialist

You should have clear goals before tasking the life cycle cost specialist. Ideally, prioritize the areas most likely to have cost savings and have studies performed early. Examples of trade-offs are:

- System maintenance concepts.
- Alternative system/product design configurations.
- Alternative production approaches, such as subcontracting.
- Alternative product distribution methods.

Three general life cycle cost analysis task categories are:

1. Baseline analysis—establishes a baseline for future cost tracking, and identifies hardware or software elements that are major cost drivers..
2. Trade study analysis—evaluates alternatives.
3. Tracking analysis—monitors variations relative to a baseline estimate.

Figure 4-3 shows the basic process of the life cycle cost analysis. TABLE 4-2 lists step-by-step procedures for an analysis.

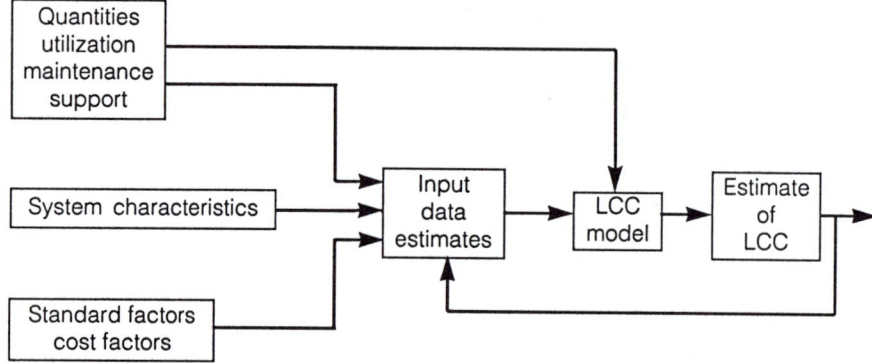

Fig. 4-3. Simplified life cycle cost analysis.

Table 4-2. Life Cycle Cost Analysis Procedures.

State objectives
Define assumptions
Select cost elements
Develop CERs
Collect data
Estimate element costs
Perform sensitivity analysis
Perform uncertainty analysis
Check assumptions
Present results

Source: *Life Cycle Cost in Navy Acquisitions*, MIL-HDBK-259.

Important tasks you may assign the life cycle cost specialist are:

- Develop or modify the life cycle cost model.
- Identify life cycle cost drivers.
- Perform sensitivity performance/cost trade studies.
- Identify optimum reliability and maintainability levels.
- Compare alternative maintenance concepts.
- Rank the design and logistical cost drivers.
- Define design-to-cost requirements.
- Perform cost of ownership assessment for operation and support.
- Conduct comparison cost studies of systems and components.
- Analyze warranty implications.
- Establish cost-of-change review guidelines.

Controlling life cycle cost

The objectives of managing the life cycle cost effort are to:

- Establish cost control.
- Organize the life cycle cost-related tasks.
- Develop a methodology that supports the understanding of requirements.
- Provide information quickly and accurately.
- Place life cycle cost tolls in the hands of the designers.

A life cycle cost plan controls the effort. The purpose of the plan is to define the approach for using life cycle costing to influence management, design, production, operation, maintenance, and support decisions. It explains how life cycle costing interacts with, and supports, design, specialty engineering, and systems engineering. It describes the:

- Responsibility for conducting LCC tasks.
- Schedule and interfacing milestones.
- Specific tasks and efforts required.
- Data flow and interfaces to related efforts such as:
    - ~ Design to cost
    - ~ Reliability
    - ~ Maintainability
    - ~ Logistic support analysis
    - ~ Integrated logistics support
    - ~ Training
    - ~ Support and test equipment
    - ~ Manufacturing
    - ~ Suppliers

The benefits of life cycle cost analyses are greatest when performed early in the project. However, the data is usually at a minimum. It is important that the data generation and communication procedures are decided early. You will likely form a team for life cycle cost. Members might include:

- The life cycle cost analyst.
- Design engineers.
- Manufacturing.
- Specialty engineering.
- Systems engineering.
- Cost estimating.

The team ranks design and logistical cost drivers for further analysis. It should also check assumptions for validity.

Summary

- Life cycle cost is the total cost to the customer of acquiring and owning a system over its full life.
- Life cycle cost allows decisions for:
  ~ Lowest cost.
  ~ Sensitivities between cost and parameters.
  ~ Trade-offs in the requirements process.
- Life cycle cost estimates are not good for budgeting directly but they do provide a basis for what is reasonable. Life cycle cost estimates are better for comparisons than absolutes.
- Early project decisions have a heavier impact on the ultimate life cycle cost of a system than those made later.

Further reading

Jones, James V. *Engineering Design: Reliability, Maintainability, and Testability*. Blue Ridge Summit, PA: TAB Books, 1988.

SAE G-11 RMS Committee. *RMS Reliability, Maintainability & Supportability Guidebook*. Warrendale, PA: Society of Automotive Engineers, Inc., 1990.

U.S. Department of Defense. *Electronic Reliability Design Handbook*. Military Handbook 338. Washington, D.C.: U.S. Department of Defense.

_____. *Life Cycle Cost in Navy Acquisitions*. Military Handbook 259. Washington, D.C.: U.S. Department of Defense.

5
Supportability

Life cycle cost must consider the maintenance and support of a system. You should design the system, its support, and its means of production concurrently. Too many products are designed with support and production as only afterthoughts. Later chapters will expand the related topics of reliability, maintainability, testability, and human engineering. In this chapter, you'll learn:

- What tasks the supportability specialist performs.
- How other specialty engineering affects supportability.
- How to control supportability.

Supportability is important because:

- Funds are limited to maintain and support systems.
- Downtime can be minimized.
- The system can be integrated with its facilities and support equipment.

Defining supportability

Supportability is the degree to which system design characteristics and planned logistics resources meet system utilization requirements. Integrated Logistics Support (ILS) as defined by the Department of Defense is a disciplined, unified, and iterative approach to the management and technical activities necessary to:

- Integrate support considerations into system and equipment design.
- Develop support requirements that are related consistently to readiness objectives, to design, and to each other.
- Acquire the required support.
- Provide the required support during the operational phase at minimum cost.

Specialty engineering

The goals of the supportability specialist are:

- To have supportability influence the design.
- Translate availability and readiness requirements into supportability requirements.
- Identify and plan for the necessary support.
- Provide support at minimum cost.

Systems engineering and supportability specialists define the supportability objectives, which makes them system requirements. Issues that can drive objectives are:

- Maintenance personnel availability or work-hour constraints.
- Personnel skill level constraints.
- Operating and support cost constraints.
- Target failures correctable at each maintenance level.
- Allowable downtime at the customer's location.
- Turnaround time to fix and maintain the system.
- Standardization requirements.

Table 5-1. Considerations for Developing Supportability Objectives.

<u>System mission requirements</u>
Operational concept
Operational environment
Utilization
Use frequency and duration
Support concept
Performance requirements
Measures for effectiveness
Mobility requirements
Service lifetime
Personnel

<u>Technology opportunities</u>
Materials
Computer capabilities
Manufacturing technology
Training devices and simulators
Built-in-test
Support equipment

<u>Logistic constraints</u>
Support money
Existing support structure
Personnel
Training
Standardization
Consumables
Access to the system
Safety and environmental concerns

Adapted from: Integrated Logistics Support Guide, Defense Systems Management College, 1986.

TABLE 5-1 shows some of the considerations for developing supportability objectives. The elements of supportability are:

- **Maintenance planning**—Formulating and establishing maintenance concepts and requirements for the system.
- **Manpower and personnel**—Identifying and acquiring the people with the skills needed to operate and support the system.
- **Supply support**—All management actions, procedures, and techniques used to determine requirements to acquire, catalog, receive, store, transfer, issue, and dispose of supplies.
- **Support equipment**—All equipment required to support the operation and maintenance of the system.
- **Technical data**—Recorded information of a scientific or technical nature.
- **Training and training support**—The processes, procedures, techniques, training devices, and equipment used to train personnel to operate and support the system.
- **Computer resources support**—The facilities, hardware, software, documentation, and people needed to operate and support embedded computers and processors.
- **Facilities**—Permanent or semipermanent real property assets required to support the system.
- **Packaging, handling, storage, and transportation**—The resources, processes, procedures, design considerations, and methods to ensure that all system and support items are preserved, packaged, handled, and transported properly. This includes environmental considerations and preservation requirements for short- and long-term storage and transportability.
- **Design interface**—The relationship of logistics-related design parameters to readiness and support resource requirements expressed in operational terms.

TABLE 5-2 expands some of the elements. TABLE 5-3 gives some sources for developing the ILS elements. TABLE 5-4 shows design factor examples that affect supportability.

Each of the elements is important but often neglected. Transportability is frequently overlooked by systems designers. Systems' components, after protective packing, do not always meet volume requirements for shipping. Critical component shipments have been known to jam in aircraft doors because they were too big! Systems engineering must define not only the operating environment, but also the shipping and storage environments. For example, products designed only for ground use have been known to leak or explosively decompress in unpressurized aircraft cargo holds. The shipping requirements were never established! TABLE 5-5 lists a few of the characteristics affecting transportability.

Developing supportability and ILS elements requires information from most of the engineering specialties. The systems engineer ensures communication and information flows effectively and at the proper times. TABLE 5-6 lists some of the engineering activities and related supportability tasks.

70 Specialty engineering

Table 5-2. Logistic Elements Example.

Supply support	Transportation Packaging and preservation Spares and repair parts Cataloging Prescreening
Support equipment	Handling Tools Test Safety Calibration Alignment Corrective maintenance Preventative maintenance
Training and training support	Instructors Documentation Curriculum Prime system Simulators Support equipment Training requirements Course outline Textbooks Training plan
Technical data	Spares provisioning Support equipment provisioning Operating and maintenance manuals Technical drawings Specifications Reports Calibration requirements Test plans Computer programs

Table 5-3. Sources to Develop ILS Elements.

Maintenance planning	Maintenance system Operational concepts Test, field experience, and historical data Reliability and maintainability predictions
Manpower and personnel	Reliability and maintainability predictions Test data Field data Historical data
Supply support	Reliability predictions Test data Field data Historical data
Support equipment	Lists of standard support and test equipment Standard tool specifications

Technical data	System functional requirements
	Production documentation
	Technical manual standards and specifications
	Descriptions of target audience capabilities
Training and training support	Existing personnel skill capabilities
	Existing programs of instruction
	Available training devices
Computer resources support	System functional requirements
	Test reports
	Field reports
Facilities	Facilities available
	Funding constraints
	Maintenance concept
Packaging, handling, storage, and transportion	Existing transportation systems and capabilities

Table 5-4. Examples of Design Factors Affecting Supportability.

Factor	Measure
Reliability	Mean-time-between-failures
Maintainability	Mean-time-to-repair
Maintenance burden	Maintenance man-hours per operating hour
Testability	Percent successful detection of faults
Transportability	Fits within standard shipping volume and weight

Table 5-5. Some Characteristics Affecting Transportability.

Physical	Volume
	Weight
	Center of gravity
Environmental	Temperature
	Pressure, altitude
	Humidity
	Vibration
	Shock
Hazards	Toxic substances
	Explosive atmospheres or pressures
	Flammable materials
	Corrosive materials

Table 5-6. Engineering Activities and Supportability.

Engineering Activity	Related Supportability Tasks
Reliability	Repair-level analysis Maintenance man-hour requirements Provisioning studies
Maintainability	Repair-level analysis Maintenance man-hour requirements
Life cycle cost	Logistic trade studies Repair-level analysis Provisioning studies
Human engineering	Personnel skill requirements Training and training device requirements
Safety engineering	Maintenance procedures Protective clothing and equipment
Testability	Personnel skill requirements Maintenance procedures Repair-level analysis Training Maintenance man-hour requirements

Benefiting from supportability

Benefits of supportability include:

- Better availability.
- Faster turnaround times.
- Less use of consumables.
- Less man-hours for maintenance, lower skill levels.
- Less expensive and fewer pieces of test and support equipment.
- Less costly transportation and handling equipment.
- Lower facilities cost.

Tasking the supportability specialist

The supportability specialist should participate in system analysis and trade studies throughout a system's life cycle. The supportability specialist can provide key information in customer need and use studies. Support element experience factors and objectives are examples of the information that can be used in the synthesis process.

Besides acting as an expert, the supportability specialist is expected to develop the ILS elements. Logistic Support Analysis is a tool that assists that effort. MIL-STD-1388-1A, Logistic Support Analysis, describes the requirements of an LSA program and the tasks to be performed. Tasks for any project must be tailored. TABLE 5-7 lists the tasks of MIL-STD-1388-1A. The supportability specialist tasks must support the systems engineering process. Tasks should work towards defining:

- The system's concept.
- Identifying supportability and cost drivers.

- System supportability and cost objectives.
- The support concept.
- Reliability, maintainability, testability, and human engineering goals.
- Support system optimization.
- Logistics support resource requirements.
- Supportability verification.

Table 5-7. Logistics Support Analysis Tasks.

Task Section	Task Description	Purpose
100	Program planning and control	Provides for planning and control of the LSA program.
101	Early LSA strategy	
102	LSA plan	
103	Program and design reviews	
200	Mission and support systems definition	Establishes supportability objectives, design goals, and constraints
201	Use study	
202	Mission hardware, software, and support system standardization	
203	Comparative analysis	
204	Technological opportunities	
205	Supportability and supportability-related design factors	
300	Preparation and evaluation of alternatives	Optimizes the support system for the new product by balancing cost, schedule, performance, and supportability.
301	Functional requirements identification	
302	Support system alternatives	
303	Evaluation of alternatives and trade-off analysis	
400	Determination of logistic support resource requirements	Identifies the logistic support resource requirements of the new product in its environment.
401	Task analysis	
402	Early fielding analysis	
403	Postproduction support analysis	
500	Supportability assessment	Ensures that requirements are achieved.
501	Supportability test, evaluation, and verification	

Controlling supportability

Controlling supportability can be difficult. Integrating supportability into the systems engineering process is a continuous and iterative activity. Also, few systems engineers become knowledgeable about the logistics area. When time is a constraint, you might prefer to treat supportability as a "black box," concentrating on the inputs and outputs, and letting the supportability specialist deal with the inner workings. Without knowing some of the workings, you might find it difficult to manage the interfaces among the specialty groups.

As a minimum, you should expect supportability to furnish you with criteria and constraints at the top system level. They should provide descriptors such as maintenance concepts, personnel constraints, training constraints, and repair constraints. Minimum analysis should define maintenance equipment, spares, repair parts, manuals, and transportation and handling, for example.

The supportability specialist should plan for people to do the analyses, the computer hardware and software, and models and tools to do the job. Training for the supportability staff and design engineering must be planned and budgeted.

Supportability factors must be defined. Transportation, maintenance, and environmental factors are part of the job. The specialist defines existing support capabilities such as supply support, personnel, facilities, and test equipment.

You should insist on a formal planning process. The plan should be updated to keep up with the changing and iterative nature of the system engineering process. Example topics are:

- Objectives.
- Scope.
- Reference documents.
- Maintenance concept.
- Support equipment description.
- Tasks.
- Processes such as LSA.
- Responsibilities.
- Interfaces to other specialty groups.
- Schedule.
- Staffing.
- Risk.

Summary

- Supportability is important because:
  - ~ Funds are limited to maintain and support systems.
  - ~ Downtime can be minimized.
  - ~ The system can be integrated with its facilities and support equipment.
- Supportability is the degree to which system design characteristics and planned logistics resources meet system utilization requirements.
- The goals of the supportability specialist are:
  - ~ To have supportability influence the design.
  - ~ Translate availability and readiness requirements into supportability requirements.
  - ~ Identify and plan for the necessary support.
  - ~ Provide support at minimum cost

- Benefits of supportability include:
  - ~ Better availability.
  - ~ Faster turnaround time.
  - ~ Less use of consumables.
  - ~ Less man-hours for maintenance, lower skill levels.
  - ~ Less expensive and fewer pieces of test and support equipment.
  - ~ Less costly transportation and handling equipment.
  - ~ Lower facilities cost.

Further reading

Defense Systems Management College. *Integrated Logistics Support Guide*. Fort Belvoir, VA: Defense Systems Management College, 1986.

Jones, James V. *Integrated Logistics Support Handbook*. Blue Ridge Summit, PA: TAB Books, 1987.

_____. *Logistic Support Analysis Handbook*. Blue Ridge Summit, PA: TAB Books, 1989.

SAE G-11 RMS Committee. *RMS Reliability, Maintainability & Supportability Guidebook*. Warrendale, PA: Society of Automotive Engineers, Inc., 1990.

U.S. Department of Defense. *Logistic Support Analysis*. Military Standard 1388-1A. Washington, D.C.: U.S. Department of Defense.

6

Reliability

Reliability is important to customer satisfaction and product effectiveness. In this chapter, you'll learn:

- What affects the reliability of your product.
- How to derive reliability requirements.
- How to task the reliability specialist.
- What should be in the reliability program plan.

Good engineering practice in reliability pays off in:

- Customer satisfaction.
- Product effectiveness.
- Reduced repairs.
- Higher safety.

Defining reliability

Your job, as systems engineer, is managing the technical effort. You should understand the terms used by reliability specialists and the activities that they should perform for your program or project.

Defining terms

What is reliability? What is failure to a reliability specialist? What does MTBF mean? MIL-STD-721C, *Definitions of Terms for Reliability and Maintainability*, has these definitions:

> Reliability: (1) The duration or probability of failure-free performance under stated conditions. (2) The probability that an item can perform its intended function for a specified interval under stated conditions. Failure: The event, or inoperable state, in

which any item or part of an item does not, or would not, perform as previously specified. Mean-Time-Between-Failure (MTBF): A basic measure of reliability for repairable items: The mean number of life units during which all parts of the item perform within their specified limits, during a particular measurement interval under stated conditions.

You might have noted that four elements are present in the definition of reliability. They are probability, performance, time, and conditions. Reliability is defined in terms of statistical occurrences. Failures may be predicted over averages of large numbers of units and long operating periods. A certain degree of randomness is present. "Performing its intended function" could be tied to the requirements specification(s). The specified interval, or time, identifies a life period or phase of the system. Stated conditions refer to the environments in which the system is used. These stress conditions directly affect the system reliability.

System failures

As the systems engineer, you should know likely failure modes of components and environmental stress factors. This information is important during the synthesis process and in assessing risk. Plans for reliability should be based on an understanding of failure mechanisms.

System failures can be traced to component failures. Components typically have dominant failure mechanisms peculiar to the type of component. Improved reliability results from knowing the failure mechanisms and taking proactive measures to reduce the failure rate.

Heat is a key stress factor in almost all reliability considerations. For electronic equipment, temperature is a major contributor to the failure rate. The cooler the equipment, the longer the time between failures. Heat also speeds any chemical reactions that might attack the functionality of a system. Corrosion is an example of a temperature-dependent process.

Electrical components are stressed mainly by voltage or current. Electronic components using semiconductors are susceptible to high voltages that damage the materials. Preventive measures are to limit the voltage potential, whether steady or transient. Power transistors can be damaged by excessive currents. Preventing excessive stress considers current as a critical factor in device application and circuit design.

Contamination will cause failures in some devices. Semiconductors and optics are susceptible to this. Controlling of manufacturing processes and vendor selection are critical to controlling this factor.

Environmental or system-induced stresses can lead to failures. The shock of the shipping truck backing into a dock is a mechanical stress. Vibration can cause devices and parts to break loose both externally and internally.

Effective reliability programs recognize the stress mechanisms and alleviate them. Careful analysis by reliability specialists is required to identify problem areas. Systems engineering must ensure that the effort is integrated across the design disciplines.

Reliability program activities

A good reliability program has certain primary activities that are essential. These are setting the requirements, designing for reliability, verifying reliability, controlling the design and manufacturing processes, and auditing the field data for the system.

Requirements before design is the system engineer's theme. System use, performance, and environments must be known and documented by the system engineer. The reliability specialist will base assumptions and calculations on this information. Contractual requirements and customer needs drive the desired failure rate of the system. The system engineer and the reliability specialist work out allocations and design rules before allowing design activities.

The reliability specialist cannot design reliability into the system. Only the design engineers can do that. The design engineers need allocations and guidelines to accomplish their work. The reliability specialist assists the design engineers with analyses and review of the design. But the burden for a system's reliability must rest on the design engineers' shoulders.

The completed design is verified against the reliability requirements. Verification can range from analysis, to testing in simulated environments, to full testing in field environments. It is up to the system engineer and the reliability specialist to determine the best approach for verification.

Controlling the design for reliability is not easy. Engineering changes can inadvertently affect reliability derating and margins. It is necessary that configuration control be reviewed by a reliability specialist. Changes in the manufacturing process can cause problems. Contaminants, improper procedures, new employees, or new machinery are examples of changes that might adversely affect product reliability. Workmanship problems and defective components will become the primary reliability concerns in a mature system.

Audits warn the reliability specialist of failure problems. A means of tracking in-house failures allows early detection of problems. Field data must also be analyzed for unforeseen use or environmental stress. During manufacturing, electrostatic discharge damage to electronic components might remain latent until failure months after shipment. Reliability is a continuing effort throughout the life of a product.

TABLE 6-1 summarizes many elements of a reliability program that must be tailored for application.

Benefiting from reliability

The cost of a reliability program and requirements should be balanced against the system's needs and the benefits gained. The indirect benefits of reliable systems and products should be considered. In some cases, ignoring reliability can result in significant liabilities to your company or organization.

Reliability improves other areas

Reliability effort pays off in many areas. The most obvious is in product or mission effectiveness. Effectiveness is influenced by availability, or the likelihood that a system will be ready to start when needed. Broken systems will, of course, be in repair or wait-

Table 6-1. Elements of a Reliability Program.

Specify	
Mission definition	Schedule
Failure definition	Requirements
Success definition	Vendor program (requirements)

Attain	
Design review	Vendor surveys
Failure mode and effects analysis	Screening tests
Reliability design analysis	Aging and surveillance program
Feasibility prediction	Critical and time-sensitive items
Worst-case analysis	Development
Part material and process program	Apportionment

Maintain	
Manufacturing controls and standards	Vendor program (audit and evaluation)
Configuration control	Education and training
Requalification program	Lot acceptance

Evaluate	
Failure analysis and classification	Reliability testing
Failure analysis follow-up corrective action	Field reliability program
Statistical analysis	Data collection and processing
Reliability measurement/demonstration	

Reporting	
Reliability program plan	Failure status report
Measurement/demonstration plan	Reliability status reprot

Adapted from: Electronic Industries Association. *Proceedings of the EIA Systems Effectiveness Workshop*. Chicago, Illinois, 18-20, September 1968.

ing for parts and, therefore, not available. Dependability also influences effectiveness. Once the system is going, dependability is the likelihood that it will remain fully capable of meeting all its requirements. Failures in operation cause the system to be not dependable. A good reliability program seeks to address these issues and ensure the proper product effectiveness.

Effectiveness leads to customer satisfaction. No business or organization will stay in operation for long if there are no customers. Customers today have high expectations in terms of reliable products. Television sets normally last for at least 10 years without failure. The automobile industry is finding that people will choose reliable cars and expect longer warranty periods. The U.S. taxpayer expects space vehicles and scientific packages to work without failure because repair in space is almost nonexistent. Sometimes, the minimum reliability requirements as stated by the customer must be exceeded in order to delight the customer. For some products, this is an effective market differentiator to gain market share.

Smart customers will look not only at the acquisition cost but also the cost to operate and maintain a system. Reliability has a direct effect on life cycle cost. Reliable systems cost less to maintain. This relationship should be studied before deciding the system reliability level.

Producing a system for a customer is seldom easy. Reliable systems reduce manufacturing cycle time by reducing troubleshooting, rework, repair, and retest. The time to manufacture and test should be a consideration in setting the overall reliability level. For example, some products, such as missiles, might have a very short operational time line. The manufacturing and test time line is usually much longer. Reliability affects production times and costs.

Poor reliability costs money

Poor reliability results in significant costs. The internal costs include lost factory capacity to rework and retest. Labor costs for troubleshooting, repair, and retesting are heavy in most companies. The cost of replaced parts and associated warehousing and tracking must also be added to the total.

Failures that occur outside the factory are more costly. Not only is there the cost to repair the failure, but additional outside service labor costs to maintain a service system. Direct, out-of-profit costs such as warranties and increased liability insurance result from poor reliability. Recall programs for defects are not only expensive but cause loss of customer goodwill and esteem.

The cost of finding failures earlier versus later is significant. TABLE 6-2 shows the cost of finding a single component failure at different stages of a product. The cost increases dramatically as troubleshooting complexity increases.

Table 6-2. Failure Costs Increase as Product is Assembled Electronics Parts Cost of Removal at Integration Levels (1984).

Application	Incoming Part	Removal at Printed Circuit Board	Systems Test	Field
Military	$10.50	$75.00	$180.00	$1500
Space	$22.50	$112.50	$450.00	—

Source: U.S. Department of Defense. *Electronic Reliability Design Handbook*. Military Handbook 338. Washington, D.C.: U.S. Department of Defense.

Setting requirements

Reliability is so important that the customers' requirements are not the only consideration. You have many business aspects which are affected by the reliability of your product or system. Reliability is a major product differentiator and high reliability can be a decisive factor in gaining or keeping market share.

System reliability should be analyzed for sensitivity to product costs. Reliability might cause lower costs through reduced rework and retest. You might even want higher reliability than your customer asks for sometimes.

Warranties and life cycle costs are directly affected by the reliability of a product or system. Reliability predictions and allocations are necessary data to forecast costs attributable to poor reliability. High reliability can save money paid under warranty programs. High reliability saves money over the life of a product for repair and spare parts.

Safety is also associated with reliability. A brake failure on an automobile, or an engine failure on an aircraft, affect safety. Even if this is not an expressed concern of the customer, it will be if the failure causes injury or harm. Product safety must be a consideration in the reliability planning and control.

What reliability is required by the system to perform its function in its environments over its intended life cycle? Setting reliability requirements for a system assumes that top-level definitions have been performed. These include:

- Mission profile, operational use, and utilization.
- Life cycle time span.
- Definition of all environments with associated time lines.

These are some typical issues you might address in developing the system reliability requirements:

- Is the mission reliability requirement satisfied?
- Are the logistics support requirements compatible with the reliability requirements?
- Are the reliability requirements consistent with the technology and manufacturing processes?
- Are the reliability requirements consistent with the system constraints?

Meaningful reliability requirements must state:

- Measurable, realistic reliability needs.
- Performance criteria for the system.
- Definition of failure.
- Conditions of use and environments.
- Means of verification.
- Period of time during system life.

Requirements cannot be vague. They must be achievable and stated in such a way that permits verification. The meaning of failure must be explicit. For example, are only critical functions to be regarded in the failure definition or *any* deviations from intended performance? Environments are needed for designers, test, and reliability predictions. What environments are covered under warranty if a warranty is offered? Is the reliability good for five seconds, one year, 20 years? Reliability requirements must be crystal clear to be effective. Quantified reliability requirements might be expressed as:

- Mean-Time-Between-Failures (MTBF).
- Probability of success.
- Mean life.
- Other failure rates.

TABLE 6-3 shows a typical military system mean-time-between-failures (MTBF) in hours.

Table 6-3. Typical Reliability Values.

Radar Systems	MTBF (Hours)
Ground rotating search radar	75 – 175
Large, fixed-phase array radar	3 – 6
Tactical ground mobile radar	25 – 75
Airborne fighter fire-control radar	50 – 200
Airborne search radar	300 – 500
Airborne identification radar	200 – 2,000
Airborne navigation radar	300 – 4,500
Communications Equipment	**MTBF (Hours)**
Ground radio	5,000 – 20,000
Portable ground radio	1,000 – 3,000
Airborne radio	500 – 10,000
Ground jammer	500 – 2,000
Computer Equipment	**MTBF (Hours)**
Ground computer	1,000 – 5,000
Ground monochrome display	15,000 – 25,000
Ground color display	2,500 – 7,500
Ground hard disk drive	5,000 – 20,000
Ground tape storage unit	2,500 – 5,000
Ground printer	2,000 – 8,000
Ground modem	20,000 – 50,000
Miscellaneous Equipment	**MTBF(Hours)**
Airborne countermeasures system	50 – 300
Airborne power supply	2,000 – 20,000
Ground power supply	10,000 – 50,000

Source: Systems Reliability and Engineering Division, Rome Air Development Center. *RADC Reliability Engineer's Toolkit*. Griffiss Air Force Base, NY.

Qualitative requirements support the main requirements. Qualitative requirements could include:

- Restrictions against certain part types.
- Limits on mechanical loading.
- Restrictions against small-gauge wire.
- Tie-down of cables to prevent mechanical stress.
- Allowances for thermal expansion and stresses.

These requirements can be applied across the project either by specifications or by design guidelines and handbooks.

Tasking the reliability specialist

The reliability effort is different for every project. Reliability specialists use expert knowledge to perform their assignments. The systems engineer consults with the reliability specialists and approves their efforts.

Reliability program tasks

MIL-STD-785B, *Reliability Program for Systems and Equipment Development and Production*, provides guidance on conducting a reliability program. Tasks should be selected and tailored to the goals of the project. Tasks in MIL-STD-785B are grouped into three sections:

1. Task Section 100—Program surveillance and control.
2. Task Section 200—Design and evaluation.
3. Task Section 300—Development and production testing.

Systems engineers should be familiar with the purpose of these tasks. Only you have the knowledge of the top-level requirements that you must ensure are being met. Your judgment is needed to make sure that resources are not being wasted or underutilized. The tasks are summarized here to help you understand the reliability specialists' job. The purpose of each task is taken directly from MIL-STD-785B.

Reliability program plan (task 101)

The purpose of task 101 is to develop a reliability program plan that identifies, and ties together, all program management tasks required to accomplish program requirements. A description of the subtasks can be found under the Controlling Reliability heading later in this chapter.

Monitor/control of subcontractors and suppliers (task 102)

The purpose of task 102 is to provide appropriate surveillance and management control of subcontractors/suppliers reliability programs so that timely management action can be taken as the need arises and program progress is determined.

It is critical to your reliability program that suppliers meet your requirements. This task gives you guidance on methods to ensure compliance. The task includes:

- Reliability requirements in specifications.
- Ensuring suppliers have a reliability program.
- Your participation in suppliers design reviews.
- A vigorous corrective-action effort.

Program reviews (task 103)

The purpose of task 103 is to establish a requirement for reliability program reviews at specified points in time to ensure that the reliability program is proceeding in accordance with the requirements.

Aspects to be discussed are listed for Preliminary Design Review (PDR), Critical Design Review, Reliability Program Reviews, Test Readiness Review, and Production

84   Specialty engineering

Readiness Review. PDR, for example, has listed:

- Reliability modeling.
- Reliability apportionment.
- Reliability predictions.
- Failure Modes, Effects, and Criticality Analysis (FMECA).
- Reliability content of specifications.
- Design guideline criteria.
- Parts program progress.

Failure reporting, analysis, and corrective action system (FRACAS) (task 104)

The purpose of task 104 is to establish a closed-loop failure reporting system, establishing procedures for analyzing failures to determine cause, and documenting corrective action. There must be working devices before this task can be accomplished. This task is normally done along with tasks in the 300 series. The failure reporting system is essential to correcting both supplier and engineering problems.

Failure review board (FRB) (task 105)

The purpose of task 105 is to establish a failure review board to review failure trends, significant failures, corrective action status, and to ensure that adequate corrective actions are taken in a timely manner and recorded during the development and production phases of the program.

This task normally works with task 104 (FRACAS). The failure review board reviews the failure data and ensures that proper corrective action is taken. Members of the board might include representatives from design, reliability, safety, maintainability, manufacturing, and quality assurance. You should decide whether or not systems engineering is to be represented.

Reliability modeling (task 201)

The purpose of task 201 is to develop a reliability model for making numerical apportionments and estimates to evaluate system reliability. The model method must be chosen first. Predictions are made using limited data and past experience. The model helps in the apportionment (allocation) process. As the product design matures, new data is fed into the model. The results are checked for compatibility with the system requirements and the allocated requirements.

Reliability allocations (task 202)

The purpose of task 202 is to ensure that once quantitative system requirements have been determined, they are allocated or apportioned to lower levels. The allocated requirements are passed down to the design engineers and suppliers. The reliability model of task 201 is used in the allocation process.

Reliability predictions (task 203)

The purpose of task 203 is to estimate the basic reliability and mission reliability of the system and to make a determination of whether these reliability requirements can be achieved with the proposed design.

Predictions provide a basis for life cycle cost and logistics support analysis. They are used in the analysis of mission effectiveness. Predictions should be made for each mode of operation and each environment.

A mission profile or user profile is necessary for accurate predictions. You should ensure that reliability specialists have accurate documentation for the profile.

Predictions will use the model selected in task 201. The model typically predicts from a projected generic part count or from a detailed stress analysis of the design components. The part count is generally used to develop initial predictions when detailed design information is not available. The detailed stress analysis is used to check the design on paper before building the product.

Failure modes, effects, and criticality analysis (FMECA) (task 204)

The purpose of task 204 is to identify potential design weaknesses through systematic, documented consideration of the following: all likely ways in which a component or equipment can fail; causes for each mode; and the effects of each failure (which may be different for each mission phase).

The FMECA results are inputs to design trade-offs, safety engineering, maintainability, maintenance engineering, logistic support analysis, test equipment design, human engineering design, and other activities. FMECA information includes:

- The effects of failure on mission capability.
- The dominant failure modes for each part.
- Safety hazard identification.
- Fault isolation procedures.

Sneak circuit analysis (SCA) (task 205)

The purpose of task 205 is to identify latent paths that cause unwanted functions that inhibit desired functions, assuming all components are functioning properly. Sneak circuits in electrical and electronic equipment is not unknown. Relays connected in parallel can interact by exchanging stored energy in their coils at power-up and power-down. The result is chattering relays, which seem to have a life of their own.

The fix is to isolate them from each other by means of diodes. An analysis can only be performed on a finished design, schematic-by-schematic. The analysis can be expensive because of the labor time but it is less expensive to fix in the paper stage before hardware has been built. The analysis is normally limited to critical circuitry.

Electronic parts/circuits tolerance analysis (task 206)

The purpose of task 206 is to examine the effects of parts/circuits electrical tolerances and parasitic parameters over the range of specified operating temperatures. Compo-

86 Specialty engineering

nent value tolerance, aging, and temperature-related value changes are evaluated in this task. This is to ensure that circuitry performs within the specification, given reasonable combinations of parts tolerance and variation build-up. Input signal variations and impedances are also included in the analysis.

This task is usually limited to critical circuits. The analysis is complex and requires the use of a computer for a reasonable task time. As more engineers have access to computer-aided design, however, it will become simpler for the computer to do the analysis as the design is done.

Parts program (task 207)

The purpose of task 207 is to control the selection and use of standard and nonstandard parts. It is important for defense systems to have parts that meet quality and reliability requirements. A parts program keeps unqualified parts from being used.

For all manufacturers, defense or otherwise, standardization makes sense. Standardization of parts means less documentation costs, lower parts cost through higher order quantities, and proven reliability on the fewer standard parts. In addition, the number of suppliers is reduced and the space needed to warehouse production and spare parts. A disadvantage is that it impedes designer's freedom to select parts for optimum performance and might discourage improvement efforts.

Reliability-critical items (task 208)

The purpose of task 208 is to identify and control those items that require "special attention" because of complexity, application of advanced state-of-the-art techniques, the impact of potential failure on safety, readiness, mission success, or demand for maintenance/logistics support. Other criteria for selecting an item as a reliability-critical item are:

- The sole failure of the item causes system failure.
- The item is known to require special handling, transportation, storage, or test precautions.
- The item is difficult to procure or manufacture.
- History has seen deficiencies.
- There is no history for confidence in its reliability.

The greatest impact on total system reliability can result from identifying and controlling a few critical items.

Effects of functional testing, storage, handling, packaging, transportation, and maintenance (task 209)

The purpose of task 209 is to determine the effects of storage, handling, packaging, transportation, maintenance, and repeated exposure to functional testing on hardware reliability. The results of this effort is to identify special procedures for maintenance and periodic field inspection. The information is also used to support design tradeoffs, definition of allowable test exposures, and other retest decisions.

Environmental stress screening (ESS) (task 301)

The purpose of task 301 is to establish and implement environmental stress screening procedures so that early failure(s) due to weak parts, workmanship defects, and other nonconformance anomalies can be identified and removed from the equipment.

Environmental stress screening is an effective way to prevent failures from reaching the customer. ESS stimulates failures before the product is shipped. The testing is nondestructive and virtually eliminates early field failures. ESS is also known as burn-in.

A detailed test plan is prepared that states the stress types and levels. Usually, the system or product is cycled through its modes while vibration and temperature extremes are applied. The test duration depends on the desired confidence of failure scrubbing and the complexity of the system. It can be beneficial on large systems to conduct ESS on selected subsystems or components before system integration and test. Problems are less expensive the earlier they can be identified and corrected.

Reliability development/growth test (RDGT) program (task 302)

The purpose of task 302 is to conduct prequalification testing (also known as Test, Analyze, and Fix, or TAAF) to provide a basis for resolving the majority of reliability problems early in the development phase and incorporating corrective action to preclude recurrence prior to the start of production.

This testing verifies the reliability predictions and allocations. It is performed on prototype or limited production hardware. While expensive to find problems in the hardware stage, it saves money later in production by identifying problems not caught in analysis. This often happens due to defective lots of components or poor manufacturing processes and workmanship.

Reliability qualification test (RQT) program (task 303)

The purpose of task 303 is to determine that the specified reliability requirements have been achieved. RQT differs from RDGT in that RQT is intended to independently prove the design rather than to make the equipment fail to fix it. It is a pass/fail test.

Production reliability acceptance test (PRAT) (task 304)

The purpose of task 304 is to ensure that the reliability of the hardware is not degraded as the result of changes in tooling, processes, work flow, design, parts quality, or other characteristics. This task is normally done after production has started. It is a continuing check on the reliability of the finished product.

MIL-STD-785B summary

The tasks summarized from MIL-STD-785B are for you to have some idea of what should be expected from reliability specialist engineering. The tasks are to be tailored according to what makes sense rather than specifying all of them or none of them indiscriminately. Because many of the tasks affect integrated logistics support and safety, those specialists' advice should be sought in the planning process. The tasks in MIL-STD-785B by themselves are not sufficient to ensure a reliable product.

Guiding the design

A possible pitfall of a strong reliability specialist function is that everyone will assume that the reliability specialists are responsible for reliability when, in fact, reliability is designed-in by the design engineer. Manufacturing can produce reliable products only to the extent that the processes have been designed.

Design engineers need a project-specific training program that communicates the design guidelines necessary to meet the system requirements. A project-specific guidebook for easy reference for parts derating and part selection is a start. New engineers need an understanding of the system environment and how to design for that environment. To ensure understanding, design engineers must be trained by the reliability specialist or a competent instructor.

There must be strong participation and expectation by systems engineering and management in a reliability program. Good training for manufacturing is needed to reduce or eliminate workmanship errors. Parts handling, cleanliness, and process control are essential. Integration with quality training is desirable.

The reliability specialist must have a derating plan for the product. Derating is operating an item at less severe stresses than those for which it is rated. Derating is effective because the failure rate of most parts tends to decrease with decreased stress levels. The failure rate model of most parts is stress and temperature-dependent. Factors to consider in determining the level of derating include:

- Reliability challenge—proven design or new concept?
- System repair—quick repair or nonaccessible?
- Safety—routine or life threatening?
- Life cycle—economical repair or complete replacement?

Tasking design engineers

Design engineers are crucial to reliable products. Your job, as the systems engineer, is to check to see that they are following engineering processes. Specifically, derating, design margins, documentation, allocation, and design reviews are areas for your attention.

- Do the engineers have derating rules and know how to apply them? Are mechanical engineers included in this effort?
- Are design margins explained and maintained? Are software engineers given design rules and margins for error?
- Are the designs documented for reliability predictions and calculations?
- Have all design engineers received allocations for reliability, do they understand them, and are they meeting them?

- Do individual design engineers present reliability as part of their design reviews, showing allocated and predicted reliability levels?
- If the process falls apart at the design stage, then you will fail to achieve conformance to requirements regardless of the amount of reliability planning and programs.

Table 6-4. Partial Electronics Reliability Design Checklist.

Management	
Permanent reliability staff?	_____
Experienced reliability engineers?	_____
Does reliability group review all drawings and specifications for adequacy of reliability requirements?	_____
Does reliability engineer/group review purchase orders and specifications to ensure that all parts and subassemblies are procured with adequate reliability requirements?	_____
Does reliability group have membership and a voice in decisions for the following:	
1. Material Review Board?	_____
2. Failure Review Board?	_____
3. Engineering Change Board?	_____
Is reliability group represented on surveys and quality audits of potential subcontractors?	_____
Does a reliability group member monitor/witness subcontractor tests?	_____
Does reliability group contain experts in the fields of components/failure analyses?	_____
Thermal	
Have detailed thermal analyses been performed to determine component/module ambient operating temperature?	_____
Are internal cooling equipment considerations sufficient to limit internal temperature rises to 20°C?	_____
Have circuit performance tests been conducted at high and low temperature extremes to ensure circuit stability over the required operating temperature range?	_____
Do potting, encapsulation, and conformal coating materials have good thermal conducting properties?	_____
Have differences in thermal expansion of interfacing materials been considered?	_____
Are components sensitive to heat located away from heat flow paths, power supplies, and other high-power dissipation components?	_____
Do components mounted on printed circuit boards (PCBs) have adequate lead lengths, and are the leads formed to relieve lead stress during thermal expansion and contraction?	_____

Table 6-4. Continued.

Vibration/Shock/Structural Requirements

Has analysis been performed to determine resonant frequencies that will be experienced in the equipment environment? _____

Are large components (over 1/2 ounce) being clamped or tied down to the chassis or printed circuit boards to prevent high stress fatigue failures or electrical leads? _____

Are cables/harnesses clamped closed to terminal connections to avoid resonances and prevent stress and failure at the point of connection? _____

All component leads have minimum-bend radii to avoid over-stressing? _____

Miscellaneous Requirements

Has consideration been given to avoid the use of dissimilar metals? _____

PCB conductor width is sufficient to handle maximum current flow without harmful heat generation or resistance drop? _____

Are PCBs uniformly coated? _____

Have worst-case analyses of statistical variation of parameters been conducted to determine required component electrical tolerances considering:

1. Manufacturing tolerances?
2. Tolerances due to temperature changes?
3. Tolerances due to aging?
4. Tolerances due to high frequency or other operating constraints?

_____
_____
_____
_____

Have considerations been given to preclude damage due to:

1. Installation?
2. Handling?
3. Transportation?
4. Storage?
5. Shelf life?
6. Packaging?
7. Maintenance environment?
8. Other environments?
   Humidity?
   Fungus?
   Sand and dust?
   Salt atmosphere?

_____
_____
_____
_____
_____
_____
_____
_____
_____
_____
_____
_____

Have parts been reviewed for proper application, have part stresses been calculated or measured, and do they meet:

1. Derating guidelines?
2. Application guidelines?

_____
_____

Do all parts selected meet the life requirements of the system? _____

Are handling requirements specified for critical and delicate parts susceptible to damage, degradation, or contamination from shock, vibration, static electric discharge, uncleanliness, etc.? _____

Table 6-4. Continued.

Have the following reliability analyses been performed:
1. Reliability mathematical models? _____
2. Reliability apportionments? _____
3. Reliability predictions? _____
4. Failure modes and effects analyses? _____
5. Criticality analyses? _____
6. Circuit analysis (nominal and worst cases)? _____
7. Thermal analysis? _____
8. Sneak circuit analysis? _____

This is only a partial list. A reliability engineering specialist should create a full list for your particular project and application.

Adapted from: U.S. Department of Defense. *Electronic Reliability Design Handbook*. Military Handbook 338. Washington, D.C.: U.S. Department of Defense.

Controlling reliability

The Reliability Program Plan controls the systematic approach to reliability. The systems engineer works closely with reliability engineering specialists on this important management document. It should be prepared by the reliability specialists, with approval by systems engineering.

As the systems engineer, you should use the Reliability Program Plan to:.

- Assist in managing an effective reliability program.
- Evaluate the program's understanding and execution of reliability tasks.
- Evaluate planning for procedures for implementing and controlling reliability tasks.
- Evaluate the adequacy of the staffing and organization of the reliability specialists.

MIL-STD-785B, *Reliability Program for Systems and Equipment Development and Production*, provides general requirements and specific tasks for reliability programs and provides guidelines for preparing and implementing a Reliability Program Plan. The standard states that "effective reliability programs must be tailored to fit program needs and constraints, including life cycle costs." Revision B emphasizes prevention, detection, and correction of design deficiencies, weak parts, and workmanship defects.

Measures of basic reliability include all item life units and all failures. It is recognized that reliability ties to operational readiness, demand for maintenance, and demand for logistic support. Regardless of the use of your product, defense or commercial, the Reliability Program Plan described in MIL-STD-785B is a good starting point for a tailored plan. Here is what task 101 of MIL-STD-785B says about the Reliability Program Plan:

Purpose. The purpose of task 101 is to develop a reliability program plan which identifies, and ties together, all program management tasks required to accomplish program requirements.

92 Specialty engineering

Task Description

A reliability program plan shall be prepared and shall include, but not be limited to, the following:

- A description of how the reliability program will be conducted to meet the requirements of the SOW.
- A detailed description of how each specified reliability accounting and engineering design task(s) will be performed or complied with.
- The procedures (wherever existing procedures are applicable) to evaluate the status and control of each task, and identification of the organizational unit with the authority and responsibility for executing each task.
- Description of interrelationships of reliability tasks and activities and description of how reliability tasks will interface with other system oriented tasks. The description shall specifically include the procedures to be employed which assure that applicable reliability data derived from, and traceable to, the reliability tasks specific are integrated into the Logistic Support Analysis Program (LSAP) and reported on appropriate Logistic Support Analysis Records (LSAR).
- A schedule with estimated start and completion points for each reliability program activity or task.
- The identification of known reliability problems to be solved, an assessment of the impact of these problems on meeting specified requirements, and the proposed solutions of the proposed plan to solve these problems.
- The procedures or methods (if procedures do not exist) for recording the status of actions to resolve problems.
- The designation of reliability milestones (includes design and test).
- The method by which the reliability requirements are disseminated to designers and associated personnel.
- Identification of key personnel for managing the reliability program.
- Description of the management structure, including interrelationship between line, service, staff, and policy organizations.
- Statement of what source of reliability design guidelines of reliability design review checklist will be utilized.
- Description of how reliability contributes to the total design, and the level of authority and constraints on this engineering discipline.
- Identification of inputs that the contractor needs from operation and support experience with a predecessor item or items. Inputs should include measured basic reliability and mission reliability values, measured environmental stresses, typical failure modes, and critical failure modes.

   The contractor can propose additional tasks or modifications with supporting rationale.

You should regard the Program Reliability Plan as a tool and not as a data item to be filed away. It is the plan for controlling and implementing reliability. It also allows you to assess the resources allocation for accomplishing the reliability effort for the product. TABLE 6-5 is a checklist for the major elements of a Reliability Program Plan.

Table 6-5. Major Elements of the Reliability Program Plan.

- Preliminary system-level reliability requirements.
- Description of program tasks.
- Control procedures and responsibility assignment.
- Preliminary schedules, milestones, and personnel loading.
- Identification of key personnel.
- Personnel organizational structure.
- Interfaces and relationships with other personnel.
- Resource identification and acquisition.
- Risk identification and alleviation of reliability problems.
- Plan to provide program-specific reliability design guidelines, design review checklists, and training.
- Documentation procedures for problem resolution.
- Verification methods and procedures.
- Major trade studies.

Summary

- Reliability is defined as the probability that an item can perform its intended function for a specified interval under stated conditions.
- The main reliability program activities are:
  - Setting the requirements.
  - Preparing the Reliability Program Plan.
  - Designing for reliability.
  - Verification of reliability.
  - Controlling the design and manufacturing processes.
  - Auditing the field data for the system.
- Reliability affects:
  - Product effectiveness.
  - Customer satisfaction.
  - Life cycle cost.
  - Production and test.
  - Maintenance.
  - Warranties.
  - Safety.
- Reliability requirements require top-level definitions:
  - Mission profile, operational use, and utilization.
  - Life cycle timespan.
  - Definition of all environments with associated time lines.

- Meaningful reliability requirements must state:
    - Quantified, realistic reliability needs.
    - Performance criteria for the system.
    - Definition of failure.
    - Conditions of use and environments.
    - Means of verification.
    - Period of time during system life.
- MIL-STD-785B, *Reliability Program for Systems and Equipment Development and Production*, provides guidance on conducting a reliability program.
- Task 101 of MIL-STD-785B describes the Reliability Program Plan.

Further reading

Ireson, W. Grant and Clyde F. Coombs, Jr., eds. *Handbook of Reliability Engineering and Management*. New York: McGraw-Hill, 1988.

SAE G-11 RMS Committee. *RMS Reliability, Maintainability & Supportability Guidebook*. Warrendale, PA: Society of Automotive Engineers, Inc., 1990.

U.S. Department of Defense. *Reliability Program for Systems and Equipment Development and Production*. Military Standard 785. Washington, D.C.: U.S. Department of Defense.

7

Maintainability

Good engineering practice in maintainability is a key factor in product effectiveness and customer satisfaction. In this chapter, you'll learn:

- How to set maintainability requirements.
- What tasks the maintainability specialist performs.
- What should be in the maintainability program plan.

Defining maintainability

Your job as systems engineer is managing the technical effort. You should understand the terms used by maintainability specialists and the activities they should perform for your program or project.

Defining terms

What is maintainability? What is maintenance? What does the term MTTR mean? [1]Maintainability is the measure of the ability of an item to be retained in, or restored to, a specified condition when maintenance is performed by personnel having specified skill levels, using prescribed procedures and resources, at each prescribed level of maintenance and repair. Mean-time-to-repair (MTTR) is a basic measure of maintainability: The sum of corrective maintenance times at any specific level of repair, divided by the total number of failures within an item repaired at that level, during a particular interval under stated conditions.

Maintainability is a design characteristic. It is dependent upon people, resources, and planned schemes of maintenance and repair. You can specify it, control it, and measure it as a design parameter. Maintenance is a result of designed maintainability.

Another definition of maintainability is quoted in the SAE *Reliability, Maintainability & Supportability Guidebook*:

---

[1]Taken from MIL-STD-721C, *Definitions of Terms for Reliability and Maintainability*.

"Maintainability is the probability that an item will be retained in or restored to a specified condition within a given period of time, when maintenance is performed in accordance with prescribed procedures and resources."

You can see the four elements of probability, specified condition, periods of time, and procedures/resources. Probability is an element because repair times are uncertain. Specified condition refers to the mission profile and specification's for the system or product. Time is a constraint because it is a limited resource. Procedures and resources set the standards for people, methods, tools, support equipment, etc., based on the design and maintenance concepts.

Maintainability program objectives

A maintainability program has certain primary objectives. These are setting the requirements, designing for maintainability, verifying maintainability, and planning maintenance and resources.

Requirements must be established before design can begin. System use, performance, and environments must be known and documented. The maintainability specialist bases the maintenance concept and maintainability design rules on the expected customer use in the customer's environments. The environments include places of use, places of maintenance, storage, shipping, and training. Customer needs should drive the requirements. The system engineer and the maintainability specialist allocate and derive requirements and design rules before allowing design activities.

The design engineers are responsible for maintainability, not the maintainability specialist. The maintainability specialist affects the design to achieve ease of maintenance. Design guidelines and analyses identify designer items that cause lower availability or higher maintenance costs. The bottom line is that the designers choose to include or not include maintainability in the design.

The completed design is verified against the maintainability requirements. Verification can be analysis, testing in simulated environments, or full testing in actual field environments. The system engineer and the maintainability specialist should work together to determine the best approach to meet the customer's need to verify the design.

Maintenance and resource planning is a major objective. System availability is calculated using data from both reliability and maintainability. Extra systems must be in use or stored to make up for unavailable systems in repair or in maintenance. Performing maintenance requires skilled people, tools, manuals, spare parts, lubricants, etc. Costs and quantities of these resources are calculated for planning and implementing maintenance.

TABLE 7-1 summarizes many of the elements of a maintainability program that must be tailored for application.

Benefiting from maintainability

The cost of a maintainability program and requirements should be examined with consideration to the system's needs and the benefits gained. Some benefits of maintainability are customer satisfaction, system effectiveness, reduced life cycle costs, and ease of manufacturing.

Table 7-1. Elements of a Maintainability Program.

- Establish maintainability requirements.
- Prepare maintainability program plan.
- Establish maintenance concepts and maintenance plan.
- Allocate maintainability requirements.
- Establish maintainability design criteria and guidelines.
- Train design engineers in maintainability.
- Perform maintainability analysis.
- Specify and control maintainability for subcontractors and suppliers.
- Participate in design reviews.
- Verify maintainability.
- Maintain configuration control.
- Prepare maintainability status reports.

Obviously, customer satisfaction is important. Customer requirements should determine the inherent minimum maintainability requirements of any system. Customers have an expectation that their problems will be taken care of swiftly and inexpensively. Proper maintainability will meet these expectations through fast repair and maintenance, at minimum cost.

System effectiveness is related to availability and dependability. Reduced maintenance and repair times mean greater availability and dependability. Availability means that the system or product will be ready on demand. Dependability means that once use of the product or system is begun, it will continue in use. Time spent in maintenance and repair is lost time to the customer in terms of system use.

Good maintainability reduces life cycle costs. These costs include maintenance labor cost, labor hours, and training. Maintainability seeks to minimize the cost per removal of parts. Product design for maintainability reduces the cost and complexity of support equipment or removes the need for it entirely. For some systems, operating and support costs exceed the initial acquisition costs.

Maintainability supports ease of manufacturing. The cost of rework, test equipment, and repair stations are reduced by maintainability designs. Troubleshooting is less demanding and fewer hours are spent on repairs.

For products under warranty, maintainability means less money spent on products that fail and require servicing.

Setting requirements

Setting maintainability requirements for a system assumes that top-level definitions have been performed by systems engineering. These definitions include:

- Mission profile, operational use, and utilization.
- Life cycle timespan.
- Definition of all environments with associated time lines.

A maintenance concept compatible with the definitions must be defined. The maintainability requirements are made compatible with the top-level definitions and the maintenance concept.

Defining the maintenance concept

The maintenance concept is a rough guideline for developing the maintenance plan for a system or product. Most maintenance concepts are made compatible with existing maintenance organizations or procedures. Sometimes, technology changes cause maintenance concepts to change completely. An example is the change in home television sets to replaceable circuit boards. Before then, troubleshooting and replacement was at the individual component level. The new maintenance concept was widely advertised as a benefit to the consumer. Repairs could be made faster, often without removal of the set from the consumer's home. Issues to be considered in the maintenance concept include:

- Maintenance worker constraints.
- Levels of maintenance.
- Sparing plans.
- Periodic testing.
- Scheduled or preventive maintenance.
- Planned support equipment.
- Turnaround time.
- Repair versus discard.

A maintenance plan will grow from the maintenance concept. The plan should describe the tasks and requirements for maintaining the system or product, evolving from the logistic analyses and reliability predictions. A mature plan includes requirements and time frames for personnel, facilities, repair parts, tools, test equipment, publications, and technical assistance. TABLE 7-2 lists additional areas to be considered for maintenance planning.

The maintenance concept includes the doctrine of the levels of maintenance. The goal is to perform maintenance closest to the user to reduce wait time. The resources necessary to do repairs, however, are not always available at the user's site. In these cases, more than one level of maintenance is required.

Most commercial product maintenance is done at a service facility. Automobiles are an example of a one-level maintenance concept. Copy machines are typically serviced at the customer's location to minimize downtime. Military maintenance is normally multilevel. The U.S. Air Force and U.S. Navy follow a three-level maintenance plan. The levels are organizational ("O" level), intermediate ("I" level), and depot. System complexity, turnaround time, mobility, personnel skills, and test equipment capability drive the multilevel concept. Figure 7-1 shows the flow of maintenance responsibility for a three-level maintenance plan from MIL-HDBK-338.

Organizational maintenance is designed for quick turnaround time. Capabilities include periodic servicing and removal and replacement of major components. Built-in tests generally assist troubleshooting. Skill levels are lowest at this level and test equipment is not sophisticated.

Table 7-2. Maintenance Requirement Segments.

Inspection and test equipment	The test equipment used to determine performance of depot maintenance specifications and requirements.
Material quality	The quality level of parts and material used for replacement, repair, or modification.
Preshop analysis	The extent of overhaul required. Included in the analysis would be procedural instructions as well as a detailed checklist to aid in evaluting items and determining the extent of their cleaning, repair, modification, or replacement.
In-process inspection	The in-process inspection requirements, including procedural and accept/reject criteria associated with each overhaul operation, such as disassembly, cleaning, repair, replacement and modification, as applicable.
Diagnostic and automated test equipment	The diagnostic and automated equipment (such as nondestructive testing, magnetic particle, dye penetration, etc.) used to determine the adequacy of repair, overhaul, or reconditioning.
Repair	The total sequential, step-by-step instructions and specifications used for repair, replacement, reclamation, rework, or adjustment for hardware items.
Assembly	The total step-by-step instructions used to assemble the hardware item.
Calibration	The level and method of calibration for all equipment and instrumentation.
Final performance check	The techniques and methods that ensure total satisfactory performance of the hardware item in accordance with the established criteria.

Source: U.S. Department of Defense. *Electronic Reliability Design Handbook.* Military Handbook 338. Washington, D.C.: U.S. Department of Defense.

Intermediate maintenance bridges the end user and the depot. Capabilities include troubleshooting modules and electronic circuit boards. Repair is usually by removal and replacement within major components. Tools, test equipment, and personnel skills are higher than the "O" level because of increased complexity.

Depot maintenance can accomplish almost any repair. The most sophisticated tools, facilities, test equipment, repair parts, and personnel skills are found here. Turnaround time is slowest because of the repair complexity. Depots are fixed in place. The depot supports many intermediate shops and their organization-level facilities.

The U.S. Army has five levels of maintenance. These are:

1. Crew
2. Organizational
3. Direct support
4. General support
5. Depot

100 Specialty engineering

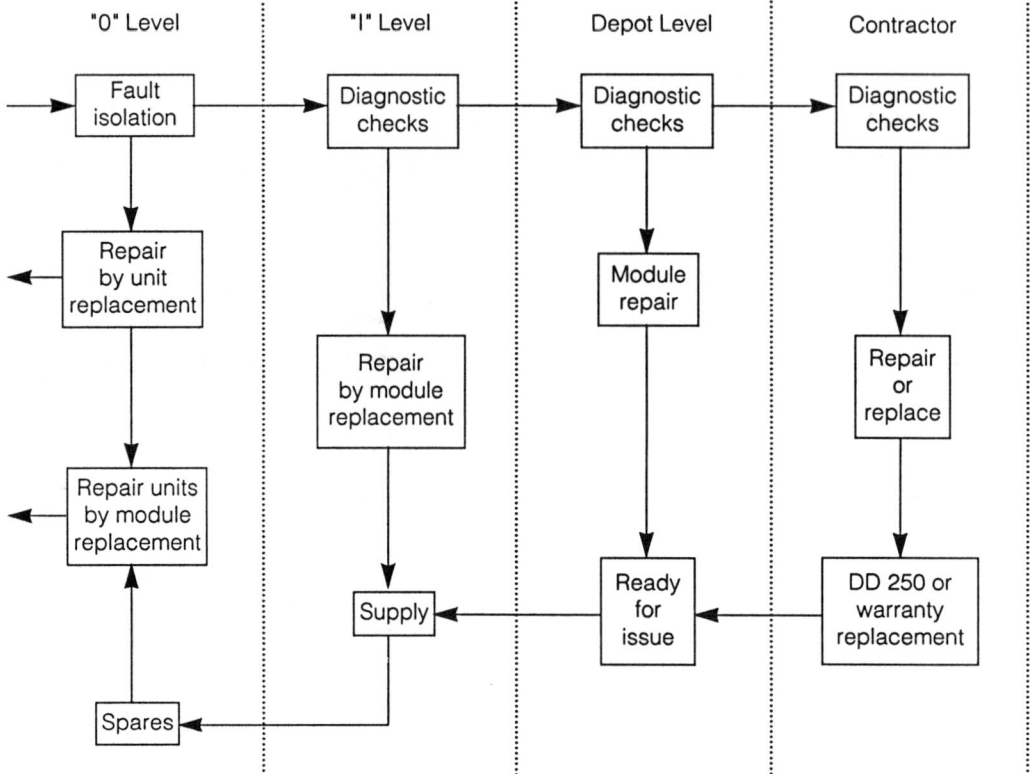

Fig. 7-1. Maintenance responsibility flowchart. (Source: MIL-HDBK-338)

Crew maintenance is performed by the crew using the equipment, such as tanks. "O" level is the same as for other services. Direct support and general support correspond to the "I" level. Direct support is mobile, which limits the test equipment and tools available. General support is less mobile and has more maintenance aids available. Depot is the same as for the other services.

Maintainability requirements

Before establishing the maintainability requirements, you must first have gathered the system information on usage, environments, and maintenance concepts. Determine the absolute "must haves" to satisfy customer needs, and formulated criteria to help you decide what is most important to your maintainability effort. The result of these efforts is detailed requirements to be given to the design engineers. What are your objectives for maintainability? The prime objectives are to minimize:

- Downtime due to maintenance.
- Cost of maintenance.
- Numbers and skill levels of personnel.

- Efforts to perform maintenance.
- Errors in maintaining systems.
- Failures induced in maintenance.

Some global issues you might want to address include:

- Is the mission maintainability requirement satisfied?
- Are the maintainability requirements compatible with logistics support, the maintenance concept, and reliability requirements?
- Are the maintenance requirements consistent with the system constraints?

Constraints might include:

- Operating hours.
- Downtime or availability.
- Mobility requirements.
- Attended/unattended operation.
- Environment (do maintainers have to work in the dark, in the cold, in hazardous conditions?).

Quantitative requirements are derived from mission requirements, customer requirements, and system-level constraints. Quantitative maintainability requirements are allocated to the system, subsystem(s), and each component. The requirements must be achievable and stated in such a way that verification is permitted. Typical requirements are mean-time-to-repair and maximum-time-to-repair at each level of maintenance, and preventive maintenance in hours per year. For example, mean-time-to-repair values usually are:

Organizational	.5 – 1.5 hours
Intermediate	.5 – 3 hours
Depot	1 – 4 hours

Qualitative requirements also support overall requirements, including interchangeability, color coding electrical hookups and fluid connectors, captive hardware, standardized connectors, common hand tools, etc. These requirements may not always be in the specifications. Sometimes they are applied through design guidelines and handbooks.

Meaningful requirements happen when system engineering has defined the top-level customer needs and system constraints. The maintenance concept must be compatible with the maintainability requirements. Design engineers need verifiable requirements allocated to their areas of responsibility.

Tasking the maintainability specialist

MIL-STD-470A, *Maintainability Program for Systems and Equipment*, provides guidance on conducting a maintainability program. MIL-STD-470A offers the following advice on tailoring its tasks.

Specialty engineering

- Every program is different.
- Every design involves compromises among different desirable characteristics.
- Programs must achieve a balance between operational need, equipment performance, cost, and schedule.
- Maintainability tasks vary from one acquisition phase to another.

Tasks in MIL-STD-470A are grouped into three sections:

1. Task Section 100—Program surveillance and control.
2. Task Section 200—Design and analysis.
3. Task Section 300—Evaluation and test.

You should be familiar with the purpose and brief description of these tasks. Your judgment is needed to ensure that the maintainability specialist is meeting the top-level requirements. The tasks are summarized here for your convenience. The purpose of each task is taken directly from MIL-STD-470A.

Maintainability program plan (task 101)

The purpose of task 101 is to develop a maintainability program plan that identifies and ties together all maintainability tasks required to accomplish program requirements.

Monitor/control of subcontractors and vendors (task 102)

The purpose of task 102 is to provide the prime contractor and the contracting activity with appropriate surveillance and management control of subcontractor's/vendor's maintainability programs. This is so that timely management action can be taken as the need arises and program progress is ascertained. This task includes specifying to suppliers information such as:

- Maintainability constraints and requirements.
- Maintenance and support concepts and requirements.
- Standardization and interchangeability requirements.
- Demonstration requirements.

Supplier programs must be compatible with your own maintainability program. You should have a representative attend suppliers' design reviews and monitor their maintainability efforts.

Program reviews (task 103)

The purpose of task 103 is to establish a requirement for the prime (or associate) contractor to conduct maintainability program reviews on scheduled dates in time to ensure that the maintainability program is proceeding in accordance with the contractual milestones and that the system, subsystem, equipment, or component maintainability requirements will be achieved.

Aspects to be discussed are listed for Preliminary Design Review (PDR), Critical Design Review, Maintainability Program Reviews, and the Test Readiness Review. PDR has listed:

- Maintainability modeling.
- Maintainability allocation.
- Maintainability predictions.
- Design guideline criteria.
- Projected personnel skill levels.
- Maintainability design approach.

Data collection, analysis, and corrective action system (task 104)

The purpose of task 104 is to establish a data collection and analysis system to aid design, identify corrective action tasks, and evaluate test results. The data collection system identifies maintainability design problems and errors and initiates corrective action. It augments predictions with preliminary trial results during design and measures demonstration results at all levels of maintenance.

Maintainability modeling (task 201)

The purpose of task 201 is to develop a maintainability model for making numerical apportionments and estimates to evaluate item maintainability when system/equipment complexity or importance warrant such a model. It is important that the model fit the level of maintenance. Maintainability mathematical models should be developed based on two things:

1. Design characteristics that affect maintainability, including:
   - Fault detection probability.
   - Frequency of failure.
   - Maintenance time.
   - Others.
2. Maintainability parameters and their relationship to system parameters:
   - Operational readiness.
   - Mission success.
   - Logistic support costs.
   - Others.

Maintainability allocations (task 202)

The purpose of task 202 is to ensure that once quantitative system requirements have been determined, they are allocated or apportioned to lower levels.

Numerical requirements must be broken down to subsystem/unit/subunit levels to establish requirements for designers. The allocations are also used to specify requirements to suppliers and subcontractors. Allocated requirements should be checked for consistency with maintainability model results.

Maintainability predictions (task 203)

The purpose of task 203 is to estimate the maintainability of the system/subsystem/equipment and to determine whether the required maintainability can be achieved with the proposed design within the prescribed support and personnel/skill requirements.

Maintainability predictions are to be made for each level of maintenance and for each subsystem and its components. Predictions should take into account the operational and support environments and maintenance concepts. The predictions are to verify that requirements will be met. Predictions can be made using MIL-HDBK-472 or other approved methods. Tasks 201 and 202 might be required to perform the predictions.

Failure modes and effects analysis (FMEA) - maintainability information (task 204)

The purpose of task 204 is to define the potential failure modes and their effects on systems, equipments, and item operation to establish necessary maintainability design characteristics, including those that must be ascribed to fault detection and isolation subsystems.

An FMEA must be prescribed in the reliability tasks for this task to happen. Uses of the information include built-in test, internal and external testers, and technical manual preparation.

Maintainability analysis (task 205)

The purpose of task 205 is to translate data from contractor's studies, engineering reports (both unique to maintainability design and developed as a consequence of other requirements), and other information available from the contracting activity into a detailed design approach. It is then used to provide inputs to the detailed maintenance and support plan, which is part of the Logistics Support Analysis (see tasks 201, 203 and 207). Elements to be considered in the analysis include:

- Mean and Maximum times to repair at each level of maintenance.
- Maintenance hours per operating hour.
- Proportion of detectable faults.
- Levels of fault isolation for built-in test and external test equipment.
- Mixes of manual, semiautomatic, and automatic tests.
- Development of unique test equipment versus existing testers.
- Design trade-offs.

Analysis should show the effects on cost and system effectiveness. The analysis is also used in the maintainability requirements allocation process. Information needed by the maintainability specialist includes:

- Operational and support concepts.
- Mission and environmental profiles.
- Overall maintainability requirements.

- Personnel constraints.
- Facility, training, skills, equipment, and tool availability.
- Lists of standard tools and equipment.

Maintainability design criteria (task 206)

The purpose of task 206 is to identify the design criteria that will be employed in translating the quantitative and qualitative maintainability requirements and anticipated operational constraints into detailed hardware designs.

Design criteria should be identified and presented at the Preliminary Design Review. It is continually refined and updated and again presented at the Critical Design Review. The information must be given to design engineers as part of the constraints and requirements for detailed design. The criteria to be considered includes:

- Design handbooks.
- Checklists.
- Accessibility.
- Interchangeability.
- Use of standards.
- Design techniques for fault detection and isolation.
- Standard tools and support equipment.
- Number of personnel and skill levels.
- Testability.
- Training needs.
- Handling, mobility, and transportability.

Preparation of inputs to the detailed maintenance plan and logistics support analysis (LSA) (task 207)

The purpose of task 207 is to identify and prepare inputs for the detailed system or equipment maintenance plan and Logistics Support Analysis. Those inputs will be based on the results of the tasks that make up the maintainability program. This is the task that affects coordination of the outputs of the maintainability program with the Logistics Support Analysis. As the inputs are prepared, consider the areas to which they will be applied:

- Skill level and personnel needs for test systems.
- Maintenance requirements at each level.
- Technical data required at each level.
- Training and equipment needed.
- Facilities required.
- Special and general purpose test equipment.

Maintainability demonstration (task 301)

The purpose of task 301 is to determine compliance with specified maintainability requirements. The maintainability demonstration is conducted according to MIL-STD-

106 **Specialty engineering**

471, *Maintainability Verification/Demonstration/Evaluation*. A maintainability demonstration plan should include the following:

- Test objectives and selection rationale.
- Identification of equipment to be tested.
- Test duration.
- Test schedule.
- Scenario for the test and personnel.
- Ground rules.
    - ~ Secondary failures.
    - ~ Technical manual usage.
    - ~ Personnel numbers and skills.
    - ~ Time limits.
    - ~ Others.

MIL-STD-470A summary

The tasks summarized from MIL-STD-470A are good starting points for a maintainability program. The tasks are to be tailored to your system's level of complexity and design goals. A qualified maintainability specialist is necessary to properly manage the maintainability effort. The tasks in MIL-STD-470A are not sufficient to ensure a maintainable product or system.

Guiding the design

Maintainability is a design task. The design engineers will make or break your maintainability objectives. They need firm requirements in terms of hard numbers and specific design rules.

Communication is necessary for selecting the solution that meets the requirements. Project-specific guidebooks and checklists show design engineers ways to achieve maintainability goals. Training by the maintainability specialist or a qualified instructor can ensure that techniques and accepted practices are understood. Design engineers need information on the user's environment and the maintenance concept to correctly design the system. Systems engineering and management must expect and ask for compliance to maintainability requirements.

Tasking design engineers

Design engineers should be following accepted practices and processes in the detailed design. Some questions you should ask design engineers are:

- Do they have design rules and guidelines and know how to apply them?
- Are the designs documented for maintainability predictions and calculations?
- Have they all received allocations for maintainability? Do they understand them? Are they meeting them?
- Is maintainability presented as part of each designers design reviews, showing allocated and predicted maintainability levels?

Table 7-3. Maintainability Design Criteria Examples.

a. To minimize downtime due to maintenance by using:
(1) A maintenance-free design.
(2) A standard and proven design and components.
(3) Simple, reliable, and durable design and components.
(4) Fail-safe features to reduce failure consequences.
(5) "Worst-case" design techniques and tolerances that allow for use and wear over an item's life.
(6) Modular design.

b. To minimize maintenance downtime by designing for rapid and positive:
(1) Prediction or detection of malfunction or degradation.
(2) Localization to the affected assembly, rack, or unit.
(3) Isolation to a replaceable or repairable module or part.
(4) Correction by replacement, adjustment, or repair.
(5) Verification of correction and serviceability.
(6) Identification of parts, test points, and connections.
(7) Calibration, adjustment, servicing, and testing.

c. To minimize maintenance costs through designs, which minimizes:
(1) Hazards to personnel and equipments.
(2) Special implements for maintenance.
(3) Requirements for depot or contractor maintenance (unless proven to be most effective).
(4) Consumption rates and costs of spares and materials.
(5) Unnecessary maintenance.
(6) Personnel skills.

d. To minimize the complexity of maintenance by designing for:
(1) Compatibility among system equipment and facilities.
(2) Standardization of design, parts, and nomenclature.
(3) Interchangeability of like components, material, and spares.
(4) Minimum maintenance tools, accessories, and equipment.
(5) Adequate accessibility, work space, and work clearances.

e. To minimize maintenance personnel requirements by designing for:
(1) Logical and sequential function and task allocations.
(2) Easy handling, mobility, transportability, and storability.
(3) Minimum numbers of personnel and maintenance specialties.
(4) Simple and valid maintenance procedures and instructions.

f. To minimize maintenance errors by designing to reduce:
(1) The likelihood of undetected failure or degradation.
(2) Maintenance waste, oversight, misuse, or abuse.
(3) Dangerous, dirty, awkward, or tedious job elements.
(4) Ambiguity in maintenance labeling or coding.

g. To minimize the frequency of tool failure by:
(1) Providing adequate accessibility, work spaces, and work clearance around fasteners to provide for solid seating of tools and for uniform application of rotational force.
(2) Ensuring torque loads required to install and remove fasteners do not exceed the capability of required tools.

System and equipment design criteria should be so structured that it enables cost-effective maintenance support throughout a deployed hardware life. This must be considered in the design process. Some examples of maintainability design criteria that might be appropriate for

Table 7-3. Continued.

some equipment programs are:

a. All repair part items having the same part numbers shall be functionally and physically interchangeable without modification or adjustment of the items or the system or equipment in which they are used.
b. Maintenance adjustment or alignment shall not be required.
c. Preventive (scheduled) maintenance requirements, including calibration, shall be eliminated.
d. Physical and functional maintenance access shall be provided to any active component upon opening or removal of access entries, and shall not require the prior removal or movement of other components.
e. Devices securing access entrances and maintenance-replaceable items shall be the captive, "quick-release" type with positive locking features.
f. Special (system or equipment peculiar) tools shall not be required in the performance of user or intermediate-level maintenance tasks.

Source: U.S. Department of Defense. *Maintainability Program for Systems and Equipment.* Military Standard 470. Washington, D.C.: U.S. Department of Defense.

Controlling maintainability

The Maintainability Program Plan controls the systematic approach to maintainability. The systems engineer works with the maintainability engineering specialists to set the requirements. It should be prepared by the maintainability specialists and approved by systems engineering. You should use the Maintainability Program Plan to:

- Assist in managing an effective maintainability program.
- Evaluate a program's understanding and execution of maintainability tasks.
- Evaluate the planning for procedures for implementing and controlling maintainability tasks.
- Evaluate the adequacy of the staffing and organization of the maintainability specialists.

The following is the Maintainability Program Plan description from MIL-STD-470A:

Purpose. The purpose of task 101 is to develop a maintainability program plan that identifies and ties together all of the maintainability tasks required to accomplish program requirements.

Task Description

A maintainability program plan shall be prepared and shall include, but shall not be limited to, the following:

- A description of how the maintainability program will be conducted to meet the requirements of the Statement of Work.
- An identification of each maintainability task to be accomplished under the maintainability program.
- A detailed description of how each maintainability task will be performed or complied with.

- The procedures (where existing procedures are applicable) to evaluate the status and control of each task.
- The identification of the organizational unit with the authority and responsibility for executing each task.
- Description of interrelationships of maintainability tasks and activities and description of how maintainability tasks will interface and be integrated with other system oriented tasks (i.e., reliability, human factors, personnel, systems life cycle and design to cost, system engineering, value engineering, safety engineering, integrated logistics support (ILS), etc.) to avoid duplication of effort. The description shall specifically include procedures to be employed which will assure that the maintainability program will operate within the goals and constraints established by front-end Logistics Support Analysis (LSA) activity, and that applicable, maintainability data derived from, and traceable to, the maintainability tasks specified are integrated into the Logistic Support Analysis Records (LSAR).
- A schedule with estimated start and completion points for each maintainability program activity or task and level of effort to be spent on each task.
- The relationship of the Maintainability Program schedule to other schedules for system engineering tasks (such as Reliability) shall be established.
- The procedures or methods for identification and resolution of problems and tracking status.
- The method by which the maintainability requirements are disseminated to associate personnel, subcontractors, vendors, and the controls levied under such circumstances.
- The identification of contractor organizational elements responsible for managing and implementing the maintainability program and a description of related management structure, including interrelationship between line, service, staff, and policy organizations.
- A statement identifying which sources of maintainability design guidelines will be utilized including all Department of Defense (DoD), internal contractor and other non-DoD prepared material.
- The procedures for recording maintainability data.

The contractor may propose additional tasks or modifications with supporting rationale for additions or modifications.

The Maintainability Program Plan is your tool, not just a piece of paper. It is your plan for controlling and implementing maintainability. It also allows you to assess the resources allocation and risk in accomplishing the maintainability effort for the system.

Summary

- Maintainability is the probability that an item will be retained in, or restored to, a specified condition within a given period of time, when maintenance is performed in accordance with prescribed procedures and resources.
- Maintenance is defined as all actions necessary for retaining an item in, or restoring it to, a specified condition.
- Mean-Time-to-Repair (MTTR) is a basic measure of maintainability.

- The main maintainability program activities are:
  - Setting the requirements.
  - Preparing the maintainability program plan.
  - Establishing maintenance concepts and plans.
  - Designing for maintainability.
  - Verification of maintainability.
- Maintainability affects:.
  - Product effectiveness.
  - Customer satisfaction.
  - Life cycle cost.
  - Production and test.
- Prime maintainability objectives are to minimize:.
  - Downtime due to maintenance.
  - Cost of maintenance.
  - Numbers and skill levels of personnel.
  - Effort to perform maintenance.
  - Errors in maintaining the systems.
  - Failures induced in maintenance.
- MIL-STD-470A, *Maintainability Program for Systems and Equipment*, provides guidance on conducting a maintainability program.
- Task 101 of MIL-STD-470A describes the maintainability program plan.

Further reading

Jones, James V. *Engineering Design: Reliability, Maintainability, and Testability*. Blue Ridge Summit, PA: TAB Books, 1988.

SAE G-11 RMS Committee. *RMS Reliability, Maintainability & Supportability Guidebook*. Warrendale, PA: Society of Automotive Engineers, Inc., 1990.

U.S. Department of Defense. *Maintainability Program for Systems and Equipment*. Military Standard 470. Washington, D.C.: U.S. Department of Defense.

8

Human engineering

Human Engineering is essential to system effectiveness. It affects personnel and training costs and helps to prevent accidents due to human error. In this chapter, you'll learn:

- How to set human engineering requirements.
- What tasks the human engineering specialist performs.
- What should be in the Human Engineering Plan.

Defining human engineering

MIL-H-46855B, *Human Engineering Requirements for Military Systems, Equipment and Facilities*, defines human engineering and human factors as:

> Human Engineering—The area of human factors which applies scientific knowledge to the design of items to achieve effective user-system integration.

> Human Factors—A body of scientific facts about human characteristics. The term covers all biomedical and psychosocial considerations; it includes, but is not limited to, principles and applications in the areas of human engineering, personnel selection, training, life support, job performance aids, and human performance evaluation.

Another term for you to know is task analysis.

> Task Analysis—A time-oriented description of personnel-equipment/software interactions brought about by and operator, controller or maintainer in accomplishing a unit of work with a system or item of equipment. It shows the sequential and simultaneous manual and intellectual activities of personnel operating, maintaining or controlling equipment, rather than a sequential operation of the equipment. It is a part of system engineering analysis where system engineering is required.

111

Benefiting from human engineering

System effectiveness is impacted by both human and machine performance. Technology advances are now leaving the person as the limiting factor. Fighter aircraft, for example, are capable of sustaining turning g-forces higher than that of a pilot's physical capabilities. Good human engineering is measured by the lack of problems when a system is used by humans. Therefore, the human and machine interface must be designed for compatibility.

Personnel and training costs can be significantly reduced by proper human engineering. The number and skills of personnel needed to build, operate, and maintain a system are a result of the design. By reducing a system's complexity early in the design phase, the number and skills of people required later will be reduced. Skills include both the operability and maintainability of the system. Training costs are directly related to the skills required. Training costs include the time needed to become qualified, instructors, media, and devices. Easy to use and maintain designs reduce people and training costs.

Human error is a primary factor in most system breakdowns. Most human error is induced by some design characteristic. Thus, proper system design can reduce human error and accidents. The behavior consequences of design decisions must be considered. For example, an automobile driver who must take his eyes from the road to adjust the radio places himself and others at risk. Understanding the payoff of human engineering and its principles are essential for systems engineers.

Setting requirements

Human engineering requirements must support the system requirements. As the system engineer, you will define the system's environment, users and maintainers, and the system's "must haves" for customer satisfaction. Human engineering objectives that support system objectives are:

- Achieving required performance by users and maintainers.
- Minimizing personnel numbers, training time, and skills needed.
- Achieving reliable people-machine combinations.
- Minimizing potential error-inducing design features.

Systems engineering must also address these questions:

- Are the human engineering requirements compatible with environments, user problems, population, reliability, the maintenance concept, and safety?
- Are the top-mission requirements satisfied?
- Are the human engineering requirements consistent with the system constraints?

Systems engineering uses functional analysis to determine requirements. In turn, functional analysis is used by the human engineering specialist. Functional allocation to person or machine determines requirements for system design. Therefore, it is essential that the human engineering specialist contribute to the system engineering process.

Time lines are derived using functional analysis. These are part of the requirements. The human engineering specialist is both a contributor and user in time line studies. The specialist contributes data on human capabilities and uses the studies for further work in work load data for manning estimates, identifying critical points of stress on people, and identifying functions incapable of being performed.

The human engineering specialist contributes to design criteria in the selection of lower-level requirements. For example, a criterion might be to have a minimum number of controls. The limits of response time to a control could be another criterion. Design criteria are a part of the synthesis process at the system and all lower levels.

Requirements might be generated by an activities analysis, which is used to define the controls and displays for a system, including the layout requirements. Activities analyses are done before the design exists. If the design already exists, then a task analysis is performed. TABLE 8-1 shows the detailed requirements and conditions affecting human performance. Others include:

- Range of acoustic noise, vibration, acceleration, shock, and impact.
- Protection from thermal, mechanical, electrical, electromagnetic, visual, and other hazards.
- Adequate space for personnel, clothing, and equipment during operation and maintenance.
- Acceptable personnel accommodations for seating, rest, etc..
- Provisions for minimizing stress effects and fatigue.
- Specific tasks.
- People and machine interfaces.
- Procedures.
- Training and experience.
- Interaction with team members.
- Management and organizational behavior.

Table 8-1. Human Engineering Documentation.

Military	
MIL-H-46855	*Human engineering requirements for military systems, equipment, and facilities*
MIL-STD-1472	*Human engineering design criteria for military systems, equipment, and facilities*
DOD-HDBK-743	*Anthropometry of U.S. military personnel*
MIL-HDBK-759	*Human factors engineering design for army material*
Commercial	
SAE J925	*Minimum access dimensions for construction and industrial machinery (Society of Automotive Engineers)*

Setting requirements is a team effort. The human engineering specialist must be competent and effective. You need the understanding from a system viewpoint of the tradeoffs involved between people and machines. Human engineering requirements are often overlooked, and leads to poor performance, increased support costs, and greater incidence of accidents.

Tasking the human engineering specialist

The human engineering specialist performs major jobs in planning and analysis. Planning is discussed in the next section. Analysis is the major topic for this section.

When looking at a shopping list of tasks, you should remember that tailoring is always necessary. Every project is different, and every design involves compromises among different desirable characteristics. You must balance performance requirements, technology, schedule, and available funds.

In general, the human engineering specialist assists in determining requirements. This includes the criteria for selecting alternative designs. The specialist determines required human performance and capabilities and adds to the constraints list. Human engineering judges the alternatives in the areas of training, manpower, and organizational impact using criteria and constraints. The results are more requirements at a lower level.

The specific analysis tasks can be found in MIL-H-46855, *Human Engineering Requirements for Military Systems, Equipment and Facilities*. The tasks are not numbered as in other standards. However, they are summarized here to give you an idea of the human engineering tasks.

Defining and allocating system functions

The defining and allocating system functions task defines and allocates system functions at the system level. The functions are analyzed within the specified and derived environments. The specialist performs functional analysis on the system, support systems, facilities, operation, maintenance, and controls, and part of the allocation is used to determine automatic versus manual implementation. The process iterates to the appropriate level of detail for the system definition.

Information flow and processing analysis

The human engineering specialist analyzes information flow and processing. Because human decisions and operations support the system objective, the specialist determines how decisions are arrived at and with what information. How the information is displayed affects the way it is processed. This task impacts the system's effectiveness and prevents accidents.

Estimates of potential operator/maintainer processing capabilities

The human engineering specialist identifies plausible human roles in the system, such as operators, maintainers, decision makers, monitors, and others. The specialist then estimates the processing capability in terms of load, accuracy, rate, and time delay for each operator or maintainer information processing function. Human engineering initially uses these estimates in allocating functions, and later in defining control, display, and communication requirements.

Allocation of functions

Human engineering participates in allocating functions. The specialist analyzes and makes trade-offs to determine which functions are better implemented by machine or human. The allocation uses operator/maintainer performance data, estimated cost data, and known constraints.

Equipment selection

Human engineering participates in the synthesis process. The specialist provides criteria used to evaluate alternative solutions at the system level. The specialist also provides constraints and "must haves" to support customer requirements.

Analysis of tasks

The human engineering specialist analyzes three groups of tasks: gross analysis of tasks, analysis of critical tasks, and work load analysis. In new designs, activities analysis replaces task analysis.

The gross analysis of tasks provides information for making design decisions. The specialist determines whether system performance requirements can be met by the combinations of equipment, software, and personnel. The analysis ensures that human performance requirements do not exceed human capabilities. The completed analysis provides basic information for manning levels, procedures, training, and logistic support analysis. It helps to identify areas that require critical human performance or areas of risk to be analyzed in detail.

Critical tasks are those most likely to have adverse effects on cost, reliability, effectiveness, or safety. For example, an unperformed critical task could lead to damage of the system resulting in a major repair or long system downtime. Analysis of critical tasks identifies the:

- Information required by the operator/maintainer, including cues for task initiation.
- Information available to the operator/maintainer.
- Evaluation process.
- Decision reached after evaluation.
- Action taken.
- Body movements required by action taken.
- Work space envelope required by action taken.
- Work space available.
- Location and condition of the work environment.
- Frequency and tolerances of action.
- Time base.
- Feedback informing operator/maintainer of the adequacy of actions taken.

- Tools and equipment required.
- Number of personnel required, and their specialty and experience.
- Job aids or references required.
- Communication required, including type of communication.
- Special hazards involved.
- Operator interaction where more than one crew member is involved.
- Operational limits of personnel performance.
- Operational limits of machine and software.

Human engineering analyzes individual and crew work load and compares it with the system's requirements. Several computer models are now available to assist this effort. Mockups and simulators provide work load information for a given scenario.

Preliminary system and subsystem design

Human engineering provides criteria, constraints, and requirements to the preliminary system and subsystem designs. The specialist embeds human engineering in the design criteria documents, specifications, and interface control documents. Human engineering requirements trace to system-level requirements.

Human engineering in equipment detail design

The human engineering specialist guides the detailed design. The specialist incorporates previous analyses, requirements, and human engineering criteria into the "design to" documents. The specialist prepares guidebooks and trains the designers in human engineering practices. Human engineering also participates in the design process both prior to, and during, design reviews. The specialist ensures requirement compliance before hardware is built and software is coded.

Studies, experiments, and laboratory tests

Human engineering conducts studies, experiments, and laboratory tests to resolve problems specific to a system. System design is a decision-making process in which requirements are defined while under time and resource constraints. It is the specialist's job to provide information that will reduce the risk of system cost and performance errors. Studies, experiments, and tests reduce this risk if they are performed in a timely manner.

Mockups and models

Mockups and models provide a basis for resolving access, work space, and other human engineering problems. Human engineering directs the construction of full-scale three-dimensional mockups when critical human performance is evaluated. When human performance measurements are necessary, functional mockups are constructed. The mockups use only the minimum essential workmanship and materials. Scale models provide an analysis tool when size becomes impractical.

Dynamic simulation

Dynamic simulation techniques are a useful human engineering design tool to predict and analyze the human-in-the-loop. An example is the flight cockpit simulator, which is used to evaluate control and display positioning. These simulators have been used to reconstruct accidents for the purposes of analyzing procedures and work loads. They also provide important design information for new aircraft concepts. You should consider dual use of simulators as design tools and as training devices.

Work environment, crew stations, and facilities design

Human engineering designs work environments, crew stations, and facilities and should consider:

- Atmospheric conditions.
- Weather, climate, and environment.
- Acceleration forces.
- Acoustic noise.
- Adequate space for personnel, their clothing, their equipment, and their movement.
- Adequate physical, visual, and auditory links between personnel and personnel and equipment.
- Provisions to minimize fatigue and stress.
- Equipment-handling provisions.
- Protection from hazards such as chemicals, electrical, and mechanical.
- Optimum illumination for visual tasks.

Equipment procedure development

Human engineering contributes to the development of procedures for operating, maintaining, and using equipment. The specialist ensures that human functions and tasks are organized and sequenced for effectiveness, safety, and reliability. The effort supports and provides inputs to logistic support analysis. Procedure development is reflected in operational, technical, and training publications.

Human engineering in test and evaluation

Human engineering participates in verifying requirements during test and evaluation. The specialist uses findings from design reviews, mockups, and dynamic simulations to guide planning for later testing. Planning directs verification that the system can be operated, controlled, maintained, and supported in its operational environment as required. Verification uses personnel who are representative of the user. This evaluation process produces quantitative measures of system performance that are functions of the human and machine interface.

Controlling human engineering

The Human Engineering Program Plan provides a systematic approach to human engineering. Human engineering prepares the plan, with approval by systems engineering. The systems engineer then uses the plan to:

- Assist in managing an effective human engineering program.
- Evaluate the project's understanding and execution of human engineering tasks.
- Evaluate the planning and procedures for implementing and controlling human engineering.
- Evaluate the adequacy of the staffing and organization of the human engineering specialists.

The plan should include:

- Objectives with measurable goals and time lines.
- The approach to be taken in meeting the requirements.
- The organization, responsibilities, and key personnel of the human engineering function.
- The resources needed to accomplish the objectives.
- An identification of the human engineering tasks to be accomplished.
- The procedures to evaluate the status and control of each task.
- A description of interrelationships of human engineering tasks and activities with reliability, life cycle cost, design-to-cost, safety, maintainability, integrated logistics support, etc.
- A schedule with estimated start and completion points for each task, and the level of effort for each task.
- Procedures and methods for identifying and resolving problems and tracking the status.
- The methods by which the human engineering requirements are disseminated to associate personnel, subcontractors, suppliers, and the controls for such circumstances.
- The procedures for recording human engineering data.

The Human Engineering Program Plan is a tool. Use it as part of your risk reduction efforts.

Summary

- Human engineering affects:
  - ~ System effectiveness.
  - ~ Personnel and training costs.
  - ~ Prevention of accidents due to human error.
- Human engineering is the area of human factors that apply scientific knowledge to the design of items to achieve effective, user-system integration.

- Human error is a primary factor in most system breakdowns. Most human error is induced by some design characteristic.
- Human engineering objectives that support system objectives include:
  ~ Achieving required performance by users and maintainers.
  ~ Minimizing personnel numbers, training time, and skills needed.
  ~ Achieving reliable people-machine combinations.

Further reading

Booher, Harold R., ed. *MANPRINT, An Approach to Systems Integration*. New York: Van Nostrand Reinhold, 1990.

Burgess, John H. *Human Factors in Industrial Design: The Designer's Companion*. Blue Ridge Summit, PA: TAB Books, 1989.

U.S. Department of Defense. *Human Engineering Design Criteria for Military Systems, Equipment and Facilities*. Military Standard 1472. Washington, D.C.: U.S. Department of Defense.

_____. *Human Engineering Requirements for Military Systems, Equipment and Facilities*. Military Specification MIL-H-46855. Washington, D.C.: U.S. Department of Defense.

_____. *Human Factors Engineering Design for Army Material*. Military Handbook 759. Washington, D.C.: U.S. Department of Defense.

9

Safety

Safety requirements are important engineering practices that impact system effectiveness, regulatory compliance, and protection from injury or damage. In this chapter, you'll learn:

- How to set safety requirements.
- What tasks the safety specialist performs.
- What should be in the Safety Control Plan.

Defining safety

MIL-STD-882B, *System Safety Program Requirements*, defines safety and risk as:

Safety. Freedom from those conditions that can cause death, injury, occupational illness, or damage to or loss of equipment or property.

Risk. An expression of the possibility of a mishap in terms of hazard severity and hazard probability.

A hazard can be defined as something that could cause injury or damage.

Benefiting from safety

A safety program reduces the costs of a product's life cycle. Besides the moral issues, costs for poor product safety include increased:

- Insurance premiums.
- Recall costs.
- Accident and claim costs.
- Management time dealing with poor safety.

The Consumer Product Safety Act was designed to protect the public against unreasonable risks of injury. The Act gave the Consumer Product Safety Commission

Safety

the authority to require a company to recall, replace, or repurchase any products it considers hazardous. The consequence of ignoring safety for many small companies is to be forced out of business. TABLE 9-1 shows how you could be found liable. TABLE 9-2 lists a few of the many safety-related laws.

The Occupational Safety and Health Administration (OSHA) regulates products in the workplace. Compliance is mandatory. In other countries, standards are dictated by other organizations (see TABLE 9-3 for examples). Your target market defines the safety requirements that must be met in all of the market countries. The "four harms" of the U.S. OSHA standards are summarized in TABLE 9-4.

Table 9-1. You Could Be Found Liable.

- Design defect (not suitable for intended use).
- Manufacturing defect (inadequate testing and inspection).
- Labeled inadequately for proper use and warnings.
- Packaging allows shipping or handling damage to cause safety problem.
- Packaging allows separation of parts or instructions causing incomplete system in dangerous configuration.
- Failure to maintain customer complaints and product failure records.
- Failure to keep records of product sales, manufacture, and distribution.

Table 9-2. Examples Of U.S. Safety Laws.

Consumer Product Safety Act
Federal Hazardous Substances Act
National Traffic and Motor Vehicle Safety Act
Occupational Safety and Health Act
Toxic Substances Control Act

Table 9-3. Safety Approvals For The Workplace (Electronic Equipment).

Country	Standards
United States	Occupational Safety and Health Administration
	Underwriter's Laboratories
Canada	Canadian Standards Association
European	International Electrotechnical Commission (IEC)

Table 9-4. Four U.S. OSHA Harms.

Electrical shock	Risk increases when circuit voltage exceeds 30V rms and 42.4V peak and when current exceeds 5 mA through a 1.5K ohm resistor.
Fire	Danger occurs when temperature exceeds prescribed limits, and in electrical circuits whose voltage capabilities exceed 42.4V and current capabilities exceed 8A peak.
Mechanical hazards	Physical danger is caused by moving parts, sharp corners and edges, and products that can tip over.
High energy levels	Risk occurs when a potential of 2V or more between adjacent parts can cause a continuous volt-ampere level exceeding 240VA, or reactive components can produce energy levels exceeding 20J.

Setting requirements

Objectives for a safety program should ensure that:

- Safety is designed into the product in a timely, cost-effective manner.
- Hazards are identified, evaluated, and eliminated.
- Historical safety data is considered and used.
- Actions taken to eliminate hazards are documented.

The precedence for conforming to system safety requirements should be to:

- Design for minimum risk.
- Incorporate safety devices.
- Provide warning devices.
- Develop procedures and training.

The goal is to design for minimum risk and eliminate problems before they occur. Safety devices can be overridden, warnings ignored, and procedures never read. Are you responsible for unintended use? You won't know until you go to court. No two people hold the same idea on what is reasonable. Even the standards do not agree.

For example, capacitors are used in high-voltage AC circuits. When the cutoff switch is opened, the capacitor could be charged to a dangerous voltage. The capacitor must have a means of bleeding off the charge to prevent accidental electrical shock to maintenance personnel. MIL-STD-454 requires that the capacitor must discharge to 30 volts or lower in two seconds. The *National Electrical Code* (a registered trademark of the National Fire Protection Association) requires discharge to 50 volts or less within one minute for capacitors rated 600 volts or less.

Some general safety requirements are:

- Eliminate hazards through design, including material selection.
- Isolate hazardous substances from people.
- Minimize hazard to people during operation and maintenance from high voltage, electromagnetic radiation, sharp edges, hot surfaces, chemicals, etc.

- Minimize risks due to environmental conditions, such as temperature, noise, vibration, etc.
- Minimize risks created by human error.
- Use interlocks and other protective devices when hazards cannot be eliminated.
- Provide distinctive markings and warnings to protect people.

TABLE 9-5 is a partial checklist for electrical system safety.

Table 9-5. Partial Safety Checklist for Electrical System.

1. Are voltages high enough to cause shock? _____
2. Are live conductors protected against accidental contact by personnel? _____
3. Are capacitors discharged to less than 30 volts within 2 seconds? _____
4. Is arcing and sparking prevented in combustible atmospheres? _____
5. Are wires and cables protected from chafing and cutting? _____
6. Are fuses and circuit breakers used to protect against overload? _____
7. Are connectors wired so that live circuits are in sockets rather than pins? _____
8. Are interlocks used to remove power during access to the equipment interior? _____
9. Are there local power cutoffs for moving and rotating machinery? _____
10. Are neutral wires unswitched? _____
11. Are switches placed on the live side of the circuit to prevent turn-on during accidental grounding? _____
12. Have currents been calculated in the neutral line for multiple switching power supplies? _____

Tasking the safety specialist

MIL-STD-882B Notice 1, *System Safety Program Requirements*, describes many of the tasks of the safety specialist. Tasks are tailored to fit the needs of the individual project. The three task sections are:

1. Task Section 100—Program management and control.
2. Task Section 200—Design and evaluation.
3. Task Section 300—Software hazard analysis.

Section 300 tasks are recommended for projects with large or complicated software packages. All tasks are briefly summarized here. Your monitoring of the safety area is needed to ensure a systems approach.

124 Specialty engineering

System safety program (task 100)

The purpose of task 100 is to conduct a basic system safety program. Task 100 consists of all the tailored tasks in the three sections at a defined acceptable level of risk.

System safety program plan (task 101)

See "System Safety Program Plan" in the next section.

Integration/management of associate contractors, subcontractors, and architect and engineering firms (task 102)

The purpose of task 102 is to provide the system integrator with management surveillance of other suppliers' safety programs. This is done by uniform analyses, risk assessment, and verification data. The method of requirements allocation for safety is addressed.

System safety program reviews (task 103)

Task 103 establishes a requirement for safety reviews and periodic reports of the status of hazard analyses, safety assessments, and other parts of the safety program.

System safety group/system safety working group support (task 104)

Task 104 sets up groups to interchange data and program status. One example would be sharing the results of mishaps and hazardous malfunctions, including recommended actions to prevent recurrences.

Hazard tracking and risk resolution (task 105)

Task 105 establishes a single, closed-loop hazard tracking system that develops procedures to document and eliminate hazards above the acceptable level.

Test and evaluation safety (task 106)

Task 106 ensures that safety is considered before and during test and evaluation. Hazards that surface during testing must be identified, documented, and worked. Test equipment must be considered in hazard analysis prior to testing.

System safety progress summary (task 107)

Task 107 provides periodic progress reports summarizing system safety management and engineering activities. It contains activities, progress, and newly recognized significant hazards.

Qualifications of key contractor system safety engineers/managers (task 108)

Task 108 establishes qualifications for key engineers and managers with coordination and approval authority. Minimum qualifications include:

- Bachelor of science degree in engineering, applied or general science, or safety or business management.
- Registration as a professional safety engineer in one of the states of the United States, or certification by the Board of Certified Safety Professionals in system safety.
- Prior experience as a full-time system safety engineer on products or systems for a minimum of three years during the preceding 10 years in defined functional areas.

Preliminary hazard list (task 201)

Task 201 compiles a preliminary hazard list shortly after the concept definition effort begins. Hazards that are inherent in design approaches are identified early for investigation.

Preliminary hazard analysis (task 202)

Task 202 performs and documents a preliminary hazard analysis. It identifies safety-critical areas, evaluates hazards, and identifies design saftey criteria. The Preliminary Hazard Analysis should be done early in concept definition so that safety considerations are included in trade-off studies and design alternatives. As a minimum, the Preliminary Hazard Analysis should consider:

- Hazardous components such as fuels, lasers, toxic substances, and pressurized subsystems.
- Safety-related interfaces such as material compatibilities, electromagnetic interference, and software/hardware.
- Environmental constraints such as electrostatic discharge, lightning, vibration, shock, fire, explosive atmospheres, and exposure to toxic substances.
- Operating, test, and maintenance procedures, including human factors engineering and system/environmental effects on human performance.
- Facilities, support equipment, and training.
- Safety-related equipment, safeguards, and possible design alternative approaches, including interlocks, barriers, and personal protective equipment.

Subsystem hazard analysis (task 203)

Task 203 performs and documents a subsystem hazard analysis. Hazards identified include component failure modes, critical human errors, and hazards resulting from functional relationships. Hazards might result from performance degradation, functional failure, or inadvertent functioning. The analysis includes:

- The effects of failure on safety, including reasonable human errors.
- The potential contribution of software events and faults.
- The determination that safety design criteria in software specifications have been satisfied.
- The determination that software changes have introduced new hazards.

System hazard analysis (task 204)

Task 204 performs and documents a system hazard analysis. The system hazard analysis determines how system operation and failure modes can affect the safety of the system. This includes critical human errors. The system hazard analysis should begin as the design matures, at about the preliminary design review, and should be updated until the design is complete. The system hazard analysis examines subsystem interfaces for:

- Compliance with safety criteria.
- Possible combinations of independent or dependent failures that can cause hazards, including failure of safety devices.
- Normal operations of systems that can degrade safety.
- Design changes that could create new hazards.
- Effects of reasonable human errors.
- Potential contribution of software.

Operating and support analysis (task 205)

Task 205 performs and documents an operating and support hazard analysis. It identifies and evaluates hazards associated with the environment, personnel, procedures, and equipment involved throughout the operation of a system. The effort should begin early enough to provide inputs to design. The analysis identifies:

- Activities occurring under hazardous conditions.
- Changes needed in requirements to eliminate hazards.
- Requirements for safety devices.
- Warnings, cautions, and special emergency procedures, including those necessitated by the failure of a software-controlled operation.
- Requirements for handling, storage, transportation, maintenance, and disposal of hazardous materials.
- Requirements of safety training and personnel certification.

Occupational health hazard assessment (task 206)

Task 206 performs and documents an occupational health hazard assessment. It identifies potentially hazardous materials or physical agents involved with the system and its logistical support. It ensures that health hazards are known so that knowledgeable decisions regarding trade-offs can be made. Areas to be considered include:

- Toxicity of materials.
- Accidental exposure potentials.
- Hazardous waste generated.
- Protective clothing/equipment needs.
- Personnel potentially at risk.
- Engineering controls that could be used such as isolation, enclosure, or barriers.

Safety verification (task 207)

Task 207 defines and performs tests and other verification methods to ensure compliance with safety requirements. During design and development, analysis identifies hazards that are removed through redesign, controls, or safety devices.

Training (task 208)

Task 208 provides training for both the developer and customer personnel who are involved in developing, testing, and operating the system. Certification is provided where needed. Training includes instruction in approved safety procedures.

Safety assessment (task 209)

Task 209 performs and documents a comprehensive evaluation of mishap risk being assumed prior to test, operation, or delivery of a system. The safety assessment identifies all safety features and procedural hazards. The report is important because it tells residual unsafe design and operating characteristics of the system.

Safety compliance assessment (task 210)

Task 210 performs and documents a safety compliance assessment to verify compliance with codes and law. It is broad in scope but general in nature. A safety compliance assessment might be the only analysis conducted on a relatively low-safety risk program. It should include:

- Identifying codes, standards, and laws with compliance verification.
- Analyzing and resolving system hazards.
- Identifying specialized safety requirements.
- Identifying hazardous materials and the precautions and procedures for safe handling.

Safety review of engineering change proposals and requests for deviation/waiver (task 211)

Task 211 performs and documents analyses of engineering change proposals and requests for deviations and waivers. This is designed to prevent additional safety hazards from being introduced while trying to correct other deficiencies.

GFE/GFP system safety analysis (task 213)

Task 213 ensures system safety analyses for government-furnished equipment or property. It is only used when equipment is provided by the customer.

Software requirements hazard analysis (task 301)

Task 301 performs and documents a software requirements hazard analysis. It examines system and software requirements and design for unsafe modes such as wrong event, inadvertent command, out-of-sequence, and inappropriate magnitude. It establishes a software safety requirements tracking system. Inputs include Task 201 and 202.

Top-level design hazard analysis (task 302)

Task 302 performs and documents a top-level design hazard analysis. It begins after the Software Specifications Review and is presented at the Preliminary Design Review. This analysis includes:

- Relating the hazards from the preliminary hazards list, the preliminary hazards analysis, and the software requirements hazard analysis to specific software modules and configuration items.
- Evaluating independence/dependence and interdependence among software elements.
- Expanding the top-level design analysis to comply with safety requirements.
- Developing design change recommendations.
- Integrating safety requirements into the Software Test Plan.

Detailed design hazard analysis (task 303)

Task 303 performs and documents a detailed design hazard analysis. Task 303 begins after the Preliminary Design Review and expands upon the top-level software hazard analysis. It analyzes the software detailed design for compliance with safety requirements. Task 303 includes:

- Relating hazards to software components defined in the detailed software design.
- Evaluating independence/dependence and interdependence of low-level software components, tables, and variables.
- Analyzing the detailed design.
- Developing design recommendations.
- Developing test requirements.
- Developing safety-related information for inclusion in documents and manuals.
- Identifying safety-critical software units to designers and providing them with safety-related requirements, procedures, and guidelines.

Code-level software hazard analysis (task 304)

Task 304 performs and documents a code-level software hazard analysis. The effort starts when coding commences and is updated throughout the system's life. Task 304 uses the results of task 303. All source code is to be documented to ensure that future changes do not introduce new safety hazards. The analysis includes:

- Safety-critical software components for correctness, completeness, input-output timing, multiple event, out-of-sequence event, failure of event, deadlocking, wrong event, inappropriate magnitude, etc.
- Implementing safety criteria.
- Possible combinations of hardware and software failures that could result in a hazard.

- Proper error-default handling.
- Fail-safe and fail-soft modes.
- Input overload and out-of-bound conditions.

Software safety testing (task 305)

Task 305 performs and documents software safety testing. This begins almost immediately after coding. It verifies that all hazards have been eliminated or controlled. Task 305 includes testing inhibits, traps, interlocks, and assertions. Both normal and abnormal condition testing are to be done.

Software/user interface analysis (task 306)

Task 306 performs and documents a software/user interface analysis. Even after safety analyses and testing, there might still be safety hazards in the system. Procedures must be developed that:

- Provide for the detection of a hazard condition.
- Control the hazard and provide a recovery methodology.
- Provide operator warning.
- Provide safe cancellation of processing or event.
- Display an unambiguous status of safety-related data and information.

Software change hazard analysis (task 307)

Task 307 performs and documents the software change hazard analysis. If changes are not analyzed, the system cannot be assumed to be safe. Analysis includes:

- Changes that affect safety.
- Ensuring the change is made properly.
- Reviewing and updating the documentation.
- Incorporating methods and procedures into the Software Configuration Management Plan.

MIL-STD-882 summary

- The tasks summarized from MIL-STD-882 are guidelines for a safety program. The safety program must be tailored. A qualified safety specialist is necessary to comply with the requirements.
- Other tasks that can be assigned the safety specialist include:
    - Preparing safety criteria and guidebooks.
    - Reviewing histories of hazards in like products or processes.
    - Assisting design engineers in meeting safety requirements.
    - Monitoring test reports and field problems for safety issues.
    - Training designers in objectives and design methodologies of safety.
    - Bringing safety issues to the systems engineer before it's too late.

Controlling safety

The System's Safety Program Plan is a part of your System Engineering Management Plan. Both plans are tailored to your project. A description of a systems safety plan is in Task 101 of MIL-STD-882. It is summarized here for you to guide the safety effort.

The plan must describe in detail tasks and activities of the system safety management and system safety engineering. As a minimum, a description of the planned approach, qualified people, authority to implement tasks, and resources needed are major topics. The staffing and organization of the safety specialists are included in the plan, and their responsibilities and relationships to the rest of the project are determined. The plan includes procedures and processes needed for safety management.

Major objectives and milestones are identified. A program schedule is made with the estimated people loading. Also, the general requirements and design criteria are listed, and hazard probability calculations and risk assessment procedures are documented. Safety training for design engineers is planned. Finally, the methods for reporting and clearing hazards are defined, as well as how safety will be verified.

Summary

- Good engineering practice in safety affects:
    - ~ Product effectiveness.
    - ~ Protection from injury or damage.
    - ~ Regulatory compliance.
- Safety is freedom from those conditions that can cause death, injury, occupational illness, or damage to, or loss of, equipment or property.
- Objectives for a safety program should include:
    - ~ Safety that is designed into the product in a timely, cost-effective manner.
    - ~ Hazards that are identified, evaluated, and eliminated.
    - ~ Historical safety data.
    - ~ Actions taken to eliminate hazards are documented.
- MIL-STD-882B Notice 1, *System Safety Program Requirements*, describes many of the tasks of the safety specialist.

Further reading

Hammer, Willie. *Product Safety Management and Engineering*. Englewood Cliffs, NJ: Prentice-Hall, 1980.

U.S. Department of Defense. *System Safety Program Requirements*. Military Standard 882. Washington, D.C.: U.S. Department of Defense.

10
Electromagnetic compatibility

Good engineering practice in electromagnetic compatibility (EMC) is important both to system functionality and to regulatory requirements. It affects product effectiveness and safety. In this chapter, you'll learn:

- How to set EMC requirements.
- What tasks the EMC specialist performs.
- What should be in the EMC Control Plan.

Defining EMC

What is electromagnetic compatibility? EMC is the ability of a system or product to function properly in its electromagnetic environment and not be a source of pollution. Susceptibility refers to a system's response to interference from unwanted energy. Emission refers to the interference from a product or environment.

You probably have experienced poor EMC at home. When the electric shaver or kitchen mixer causes snow on the television, this is poor EMC. Alternator whine in your car stereo is another example. FM radios are not allowed to be played on aircraft because their oscillators interfere with the aircraft navigation radios. Newspapers have stories of helicopters in trouble near radio towers and bombers that interfere with themselves.

Unfortunately, many products and systems are designed without consideration to EMC. Fixes are applied when testing or field use reveals functional deficiencies. Sometimes, products are kept from the market by government agencies when the manufacturer is surprised to learn of certification laws.

Your job, as the systems engineer, is to take a systems approach to EMC. EMC must be designed into the system to conform to requirements at the lowest cost. There are four general classifications for EMC:

1. Radiated emissions.
2. Conducted emissions.
3. Radiated susceptibility.
4. Conducted susceptibility.

Radiated indicates that the energy is transferred and spread out through a medium in space by an electromagnetic field. Conducted means that the energy is transferred through common connections such as grounds, signal cables, and power cables. Transfer can occur over a combination of radiated and conducted paths. For example, energy might couple onto a signal cable by radiated means and transfer into the system by conduction.

Conducted emissions and conducted susceptibility have two main routes in common. Direct connections, which can be metallic leads or common ground returns, are one type of path. The second is coupling, which can be either inductive or capacitive.

Inductive coupling occurs when a conductor is present in an electromagnetic field. The simplest example is that of an ordinary transformer. Physical geometry comes into play in the degree of coupling. Loops in cables, wiring, circuit boards, and ground loops are sources and receivers of interference. Good engineering practice dictates minimum loop areas and good ground schemes.

Capacitive coupling is the linking of one circuit with another by means of the capacitance existing between them. The degree of coupling depends on the source voltage, the frequency, and the value of the interconnecting capacitance. Robust design practices include physical separation to reduce capacitance and low impedances to minimize noise conversion.

Electromagnetic signals are classified relatively as narrowband or broadband. Narrowband signals occupy very small portions of the electromagnetic spectrum. The best example of this is the continuous sinewave, which has all its energy concentrated at a single frequency. Usually, signals that occupy only tens or hundreds of kilohertz of bandwidth such as AM, FM, and single-sideband transmitters fall into the narrowband category. Broadband signals can have energy across tens or hundreds of Megahertz or more. This type of signal results from narrow pulses having relatively short rise and fall times. Examples of these sources include electric motors, pulse radar transmitters, and electrostatic discharge. TABLE 10-1 lists of examples of both narrowband and broadband interference sources.

A special case of EMC that can cause loss is electrostatic discharge (ESD), which is particularly hazardous to electronics. The source is static electricity, which is present just about everywhere. Static electricity is caused by the contact and later separation of nonconducting materials. The charge transfer that occurs at initial contact does not equally reverse at separation. The two materials now have one item charged positively, and one negatively. The materials can be different or the same, as in the case of plastic bags.

You generate static charge when you walk across carpet. Your body conducts the charge and discharges it when it touches or comes close to doorknobs or electronic equipment with exposed metallic surfaces. Static discharges are both high voltage and

Table 10-1. Typical Man-Made Interference Sources.

Transient	Broadband Intermittent	Broadband Continuous	Narrowband Intermittent	Narrowband Continuous
Mechanical function switches	Electronic computers	Commutation noise	cw-Doppler radar	Power-line hum
Motor starters	Motor speed controls	Electric typewriters	Radio transmitters and their harmonics	Receiver local oscillators
Thermostats	Poor or loose ground connections	Ignition systems	Signal generators, oscillators and other types of test equipment	
		Arc and vapor lamps		
Timer units	Arc	Pulse generators	Transponders	
Thyratron trigger circuits	Welding equipment	Pulse radar transmitters	Diathermy equipment	
	Electric drills	Sliding contacts		
		Teletypewriter equipment		
		Voltage regulators		

Source: U.S. Naval Air Systems Command. *Electromagnetic Compatibility Design Guide for Avionics and Related Ground Support Equipment.* NAVAIR AD 1115. Washington D.C.: U.S. Naval Air Systems Command, Department of the Navy, 1975.

high current. Voltages can be 20,000 volts and currents 40 amps in the brief discharge time. Current rise time is about 1 nanosecond and fall time is about 100 nanoseconds. This means that the pulse has a very high frequency component that must be considered in design for ESD immunity.

Models for describing the human body discharge can be found in the International Electrotechnical Commission Standard 801-2, DOD-HDBK-263, and the Society of Automotive Engineers, among others. A point to be made is that there is a body of knowledge on the subject that should be used in requirements, design, and verification.

Benefiting from EMC

Products and systems benefit from good EMC practice in three main areas: systems effectiveness, safety, and legal compliance. The electromagnetic environment is a condition in which the system must perform its function. If the system cannot meet all of its requirements when in the environment, it is not fully effective. EMC seeks immunity to susceptibility so that effectiveness is not affected. Product requirements and design must take into account the household, industrial, and field environments.

Safety is directly affected by EMC. In fact, the lack of EMC, as reported by the media and industry accounts, might have contributed to several aircraft crashes. Electromagnetic emissions above certain levels are considered harmful to people, consequently the care mandated safety levels that must be complied with.

Emission levels are now controlled by law to reduce electromagnetic pollution. Products are subject to impoundment and are barred from sale if found to be out of compliance. The United States, Germany, and Japan all have EMC requirements.

Ignorance of EMC can now actually cause your company to go out of business because of the potential for product delay or government ban. Qualified EMC specialists in both prevention and testing should be included on your project team.

Setting requirements

Setting requirements for EMC demands that you know the objectives, constraints, and environments of the system. Top-level requirements for the system are set first, and then a flowdown of allocated requirements is performed. Quantitative requirements are more difficult for EMC because of changing environments, such as when physical barriers such as metallic boxes are crossed. Qualitative requirements can be set more easily. These can be standardization of electronic logic family class, use of cable shielding, standard interfaces, etc., which can be verified with a yes or no.

EMC objectives should be:

- Conformance to performance requirements under all specified EMC environments.
- Compliance with all regulatory requirements.
- Minimize the cost of conformance by preferring prevention before correction.

Constraints are numerous. Some of the standards and organizations are listed for you in TABLE 10-2.

Tasking the EMC specialist

Electromagnetic compatibility is a realm for the specialist. The legal requirements for products cause risk. The risk is the potential loss if the product cannot be sold or must be recalled. Good risk management dictates the use of the most qualified person available within reasonable expense. The EMC specialist should deal with preventing problems and verifying requirements. Ideally, they should calculate EMC parameters, provide fixes for areas out of specification, and translate test methods to easily understood requirements. There are now specialists who deal only with FCC test methods because of their complexity.

The following tasks are a guide to tasking the EMC specialist. The importance of finding the right person for this assignment cannot be overemphasized.

Table 10-2. EMC and ESD Standards and Organizations.

EMC

U.S. military

MIL-STD-461, *Electromagnetic emission and susceptibility requirements for the control of electromagnetic interference*

MIL-STD-462, *Electromagnetic interference characteristics, measurement of*

MIL-B-5087, *Bonding, electrical and lightning protection*

MIL-E-6051 Electromagnetic compatibility requirements, systems

MIL-HDK-253 Guidance for the design and test of systems protected against the effects of electromagnetic energy

Radio Technical Commission for Aeronautics

DO-160 Environmental conditions and test procedures for airborne equipment

U.S. Federal Communications Commission
Code of Federal Regulations, Title 47

Part 15 for radio frequency devices

Part 15, Subpart J which applies to digital electronics

Part 18 for industrial, scientific, and medical equipment

Part 68 for equipment connected to the telephone network

International Special Committee on Radio Interference (CISPR)

CISPR Publication 22

Society of Automotive Engineers, AE-4 Committee

American National Standards Institute (ANSI)

C95 radio frequency radiation hazards

U.S. radiation control for health and safety act of 1968

German Verbund Deutscher Elektrotechniker (VDE)

Japan Voluntary Control Council for Interference

ESD

U.S. Military

MIL-STD-1686, Electrostatic Discharge Program for Protection of Electrical and Electronic Parts, Assemblies and Equipment (Excluding Electrically Initiated Explosive Devices)

DOD-HDBK-263, Electrostatic Discharge Control Handbook for Protection of Electrical and Electronic Parts, Assemblies and Equipment (Excluding Electrically Initiated Explosive Devices)

International Electrotechnical Commission 801-2

Society of Automotive Engineers

CISPR

ANSI C63

Defining the environment

Defining the environment task defines the EMC environment in terms of system requirements. Measurements of the actual environment, handbooks, or test methods can be used to prepare environment studies. The environment should be a part of the mission or user profile for the system. EMC and ESD levels for each environment, such as use or maintenance, are specified. The definition is used in the later allocation of requirements.

Preparing the control plan

The purpose of preparing the control plan is to make a tool for planning and controlling the EMC efforts to conform to requirements. In general, the plan should address the objectives and approach to achieving those objectives. The allocation of resources, such as personnel, consultants, test facilities, and money, must be addressed. The schedule of the activities and their precedence is stated. The relationship and responsibilities of the EMC specialist and the engineering team are defined. Training and guidelines are addressed. Verification at all levels is planned and criteria for completion is established. Configuration control of changes is planned so that EMC is not degraded later. The specifics of EMC control are described later in this chapter.

Providing analysis

The purpose of providing analysis is to provide specialty expertise in solving problems needing analysis. Analysis by the EMC specialist might include the use of computers, tables, engineering models, or expert judgment. Shielding effectiveness, coupling into and out of cables, and radiated emissions are examples of analysis areas. Analysis is necessary before allocating requirements to lower levels. These tasks are needed in the synthesis stage in the evaluation and choice of design alternatives.

Identifying problem areas

The purpose of identifying problem areas is to identify problem areas for risk management. Typical problem areas are high-field environments, components susceptible to ESD, restrictions against using shielded cables and connector backshells, enclosure materials, high voltage and high frequencies, co-interference, metallic oxide rectification, and rapid rise and fall times in electrical signals. Problem areas should be identified for causes and alleviation.

Reducing risk

To reduce risk and potential losses, the EMC specialist looks at identified problem areas, history of past problems, skill and ability levels of engineering staff, and possible disruptions to the EMC effort. Causes of potential and real problems are attacked to prevent their occurrence or to alleviate their affects on the system.

Providing design guidance

Guidance and expertise must be provided to design engineers before and during detailed design. The EMC specialist looks at the requirements, the technology, and the capabilities of engineers for the project. The engineers are then given the proper level and amount of training for them to carry out their assignments. Design guidelines specific to the project are distributed. The guidelines tell the requirements in easily understandable terms and offer solutions and accepted practices. The EMC specialist is located physically with the engineers and is available for consultation and problem-solving.

Allocating requirements

Requirements must be allocated for EMC from the system level down to the individual circuit board or module level. Many engineers are fearful of specifying EMC requirements at the module and board level. This is probably because of the lack of understanding of EMC calculations. This condition is improved today with the availability of personal computers and readily available software specific to solving EMC problems. You, the systems engineer, must force the issue and mandate that EMC requirements be allocated to individual modules and boards. If the design engineers are not used to receiving EMC requirements, they will feel that the EMC specialist is overstepping the traditional bounds.

Allocation is important to the success of the project. Allocation gives the individual design engineers tailored requirements. Design without requirements is a recipe for disaster. The allocation process forces an analysis of the EMC environments outside of the system, within the system, and within the modules and circuit boards. The environment seen by a module within an enclosed metal box is not the same as seen by the outside of the system. Detailed design inside the box using system-level requirements will be in error. If the box produces interference, then matters will be worse. Most EMC disasters can be prevented before formal testing. Requirements and verification at the module and box levels will save orders of magnitudes of scheduled time and money.

Integrating system parameters

The purpose of integrating system parameters is to have a systems approach to EMC. A prime example of a topic that must be approached systematically rather than haphazardly is grounding. Grounding is essential for EMC and controlling ESD. There are also many shock and safety hazards related to grounding. Grounding must be a part of system design and must be analyzed prior to detailed design.

EMC requires a systems approach because only one problem is needed to cause nonconformance. Many engineers are baffled by EMC problems because they attempt a fix that does not improve the problem. After several attempts without improvement, they are convinced the problem cannot be solved. It can be solved only after *all* causes have been tracked down and remedied. It is somewhat like a leaking pipe in which all the leaks must be fixed before the water will stop flooding the house.

Standardizing the design

The purpose of standardizing the design is to standardize the EMC approach and design. The EMC specialist can save time and resources for the project by providing solutions to generic problems. Interfaces, shielding, and modeling data are good examples. The EMC specialist can provide design engineers with a standard electrical interface that provides the most rejection to interference. Design time and analysis is reduced across the project. The EMC specialist can calculate the required shielding and standardize box and cable design. Modeling data for calculations can be standardized and provided to the design engineers. A readily identifiable savings can be identified to justify expenditures for EMC to management.

Predicting EMC

The purpose of predicting EMC is to provide feedback on proposed designs in relation to requirements. Predicting EMC is done at several stages of the engineering process. Analytical EMC models permit the EMC specialist to:

- Establish requirements.
- Evaluate proposed designs.
- Predict conformance or nonconformance to requirements.
- Isolate problems.
- Rank problems as to seriousness.
- Evaluate proposed solutions to problems.

Reviewing the design

Designs must be reviewed for EMC. The EMC specialist should be both a presenter and a reviewer at design reviews. The system approach and requirements can best be handled by the EMC specialist. Presentations by other engineers should be examined by the EMC specialist for conformance to EMC requirements. Conflicts of requirements should be challenged for early resolution.

Verifying the design

Verifying the design ensures that all EMC and ESD requirements have been met. The EMC specialist writes the EMC Test Plan. Testing differs for military, FCC, or VDE standards. The EMC specialist advises on methods of testing and preparation for test. Resources for verification should be identified and budgeted. You should decide if the EMC specialist verifies each configuration item prior to system integration.

Tasking design engineers

Products can be brought into EMC compliance in two ways. One is to fix all problems that develop. This usually results in large expenses and market delays. The other way is to prevent problems before they occur. Knowledgeable design engineers save time and money by designing to prevent problems. Design guidelines, training, and allocated requirements allow them to do their job efficiently and effectively.

There are numerous design considerations. In general, good electrical design should use signal levels that achieve desired signal-to-noise ratios both for analog and digital design. Low impedances are desirable to minimize the effects of capacitive coupling. Power lines should be filtered, and electrical layout should be looked at from a high-frequency viewpoint and not a direct-current viewpoint.

Physical considerations are mainly enclosures, cables, and connectors. Enclosures must be electromagnetically tight and designed to alleviate the effects of ESD. Cables should be shielded and cable shields terminated in special EMI backshells to connectors.

Devices in electronic equipment should use the lowest clock speed for the design application. Slew rates of signals should minimize the time rate of change to keep emissions at lower frequencies.

Software should be robustly designed to withstand the effects of EMI, ESD, and power dropouts. Automatic recovery from faults should be designed into the software. Errors caused by low signal-to-noise ratios in the presence of interference should be considered in design. Special software for diagnostics during ESD testing should help pinpoint the area being affected by testing.

Design engineers must know how to run ESD diagnostics on prototypes. Physical areas are to be divided, marked, and malfunction criteria determined. Using an ESD gun at appropriate levels can help find design failures before shipment to customers.

Controlling EMC

The EMC Control Plan and the EMC Test Plan aid the control of EMC. These plans are briefly described in this section.

EMC control plan

The EMC Control Plan is the formal document that describes the control procedures and activities. It details the EMC documentation, requirements, specifications, test methods, and criteria. For military systems, the plan is identified both in MIL-STD-461 and MIL-E-6051. The basic outline of a plan can be:

1. Introduction.
2. Application interpretation.
3. Specification interpretation.
4. Program management requirements.
5. Design requirements.
6. Test requirements.
7. Quality assurance tests.
8. Definition of terms.

Areas specifically addressed by design requirements might include:

- Electrical bonding and grounding.
- Shielding.
- Transient control.
- Radiated signal control.
- Interference and susceptibility prediction.
- Cable considerations.

TABLE 10-3 is a checklist for your EMC Control Plan.

Table 10-3. EMC Control Plan Checklist.

Introduction

1. Is the purpose of the document specified? _____

2. Is the document applicable to all subsystems and components? _____

3. Is the document applicable to all subcontractors and to all off-the-shelf purchased parts and components? _____

Management Section

1. Is there a positive statement concerning program management's intent to fully support an effective EMC control program? _____

2. Is there a sufficient organizational presentation to ensure ready implementation of the EMC control plan? _____

3. Have all critical organizational elements, including program management, EMC engineering, testing, quality assurance, materials, publications, and system design been identified in the organization? _____

4. Does the control plan specify all applicable EMC documentation to be provided? _____

5. Has the responsibility for preparing critical EMC documentation, including control plans, procedures, test plans, and other documentation, been made? _____

6. Has responsibility for testing at subsystem and component levels as well as system levels been made? _____

7. Have appropriate distinctions been made between qualification acceptance and compliance testing? _____

8. Is an individual or group of individuals designated to provide representation on behalf of EMC during critical design reviews and preliminary design reviews? _____

9. If the representation provided above is made by a team of individuals, is at least one of the individuals an EMC engineer? _____

10. Is the organization responsible for preparing, approving, and submitting deviation requests and/or waivers against EMC requirements, limits, or specifications been identified? _____

11. Does the document provide for related status of EMC efforts in relation to critical milestones of the entire program? _____

12. Does the document show program milestones for EMC? _____

13. Does the document provide for a duly constituted EMC board that will merge management and EMC engineering organizations? _____

14. Is it clear how design testing and management organizations interface together? _____

Table 10-3. Continued.

15. Does descriptive information on the EMC organization make clear how liason among the elements are to be carried out? _____

16. Is there provision for EMC presentations to be made concurrent with preliminary design reviews and critical design reviews? _____

Applicable Documents

1. Have all applicable system documents for systems and subsystems been listed? _____

2. Have the latest or applicable version of EMC specifications been listed? _____

3. Have all applicable documents relating to reliability and quality assurance been listed? _____

4. Have all applicable specifications regarding materials and processes been indicated, including those for metal treatment, finishes, chemical films, etc.? _____

5. Have all ancillary drawings and specifications regarding grounding planes, harnessing schemes, cable routing, etc., been listed? _____

6. Where more than one specification applies, is the order of precedence indicated? _____

7. Is it clear how the control plan interfaces with applicable directives and other EMC specifications? _____

8. If separate individual specifications have been developed for system-level testing and component and subsystem level testing, have these been listed and in their correct order of precedence? _____

9. Have all special provisions for deviations from specification limits, frequency ranges, test procedures, etc., been listed? _____

Analysis and Design

1. Have key operational parameters of the overall system been adequately and concisely described? _____

2. Have the operating frequency and power output levels of all emitters and susceptible devices been delineated? _____

3. Has a general presentation been provided on potential problem areas? _____

4. Have critical potential victim elements been defined? _____

5. Has an analysis been made of the power system utilization and potential susceptibilities via the power system, including grounds? _____

6. Have harmonics been considered when providing a table for predicting interference problems? _____

7. Have potential nonlinear elements in the system been identified that could generate spurious responses? _____

Table 10-3. Continued.

8. Have basic and specific guidance been provided in the following areas:

    a. Selection of interference-free parts and components? _____

    b. Bonding and grounding? _____

    c. Wiring and cabling, including connectors, grouping, bundling, and routing? _____

    d. Shielding and filtering sensitive circuits? _____

    e. Printed circuit coupling? _____

    f. Diode and transistor logic levels? _____

    g. Filters? _____

Source: U.S. *Electromagnetic Compatibility Design Guide for Avionics and Related Ground Support Equipment.* NAVAIR AD 1115. Washington, D.C.: U.S. Naval Air Systems Command, Department of the Navy, 1975.

EMC test plan

The EMC Test Plan documents a well thought-out implementation of verification. It allows a rational allocation of resources and time to EMC testing. EMC Test Plans might contain:

- Justification for running each test, which is tied to requirements traceability. The application of test results in identifying or resolving problem areas should be discussed.
- Description of each test, including:
  - Unit to be tested.
  - Test sets.
  - Test configuration.
  - Performance criteria.
  - Level of test.
  - Format of output data.
- Description of how characteristics of simulated signals will be obtained.
- Description of procedures to be followed to ensure proper operation of product or system during testing.
- Description of test instrumentation calibration procedures to be used.

TABLE 10-4 is a checklist for preparing an EMC Test Plan.

Table 10-4. EMC Test Plan Checklist.

Introduction	
1. Is the purpose of the test indicated?	_____
2. Are applicable drawings, documents, specifications, and standards listed? Is the listing complete?	_____
3. Is the test sample or system completely identified?	_____

Table 10-4. Continued.

4. Is it clear who would perform the test and at what location? _____

Description of Test Sample

1. Is the test sample or system under test functionally and physically identified? _____
2. Is it clear whether the system under test is a production sample, prototype, or engineering model? _____
3. Are external controls, ground support equipment, or ancillary systems required for testing identified? _____
4. Is the operational configuration of the system or sample during test clearly specified? _____

Test Conditions

1. Is the test facility and test location completely identified and described? _____
2. Are the facilities for test including shielded enclosures, power line filtering, and power line load distribution completely identified? _____
3. Are the ambient conditions clearly identified? _____
4. Is the arrangement of the test system during test clearly specified? _____
5. Do the test setup drawings and specifications completely identify the power distribution to the system under test, location, and orientation of exercising equipment and placement location of breakout boxes monitoring and equipment locations? _____
6. Are external loads clearly labeled? _____
7. Is the test sample bonding in accordance with MIL-STD-461? _____
8. Are test and monitoring points appropriately listed? _____
9. Are the system operating modes clearly delineated? _____
10. Are separate tests to be performed in terms of steady state interference and susceptibility and transient performance? _____
11. If so, are these completely delineated? _____
12. Are susceptibility test modes responsive to stimuli requirements of the applicable specifications? _____
13. Are all possible susceptible parameters listed and specified? _____
14. Are the initial system or sample operating conditions and control settings listed for susceptibility and interference testing? _____
15. Are critical cables between support equipment and bench test equipment shielded in order to eliminate interaction with the test setup? _____
16. Has care been taken to ensure that the test monitoring equipment themselves are not susceptible or interfering sources? _____

Table 10-4. Continued.

EMC Test to be Performed

1. Are all tests in consonance with applicable specifications? _____
2. Are deviations and special exceptions clearly documented? _____
3. Is the listing of each test to be performed complete? _____
4. Are all legitimate operating modes of the system considered for these tests? _____
5. Is the test listing compatible with performance requirements? _____
6. Are there any tests that are not required by reason of prior qualification? If so, has the prior qualification been evaluated with regard to the applicable specification? _____
7. Are exceptions being made to the EMC Control Plan? _____
8. Are the EMC tests to be performed in complete agreement with the EMC Control Plan? _____
9. Are there any deviations from the EMC Control Plan test requirements? _____
10. Are test limits thoroughly described for interference? _____
11. Have special tests been identified in relation to unique system requirements? _____
12. Have susceptibility threshold limits been carefully interpreted? _____

Test Equipment

1. Have all of the required EMC test instrumentation been listed? _____
2. Are the support equipments for monitoring and exercising the system or sample been made complete? _____
3. Have special test interfaces of breakout boxes been identified? _____
4. Is the equipment list complete? _____
5. Does the equipment list include all applicable calibration dates? _____
6. Does the calibration conform to MIL-C-45662? _____
7. Have all special-purpose devices been identified and described, including feedthrough filters, LISN's, and voltage probes? _____
8. Does the list of equipment include all applicable operational parameters including frequency range sensitivity, etc., as applicable to the parameters to be investigated? _____
9. Is the accuracy specified? _____

Table 10-4. Continued.

Detailed Test Procedures
1. Do the test procedures conform to MIL-STD-461, MIL-STD-462, MIL-E-60510, or specific specification? _____
2. Have test frequencies been selected on the basis of real system operational requirements? _____
3. Have deviations necessitated by system operational parameters from commonly used specifications been clearly delineated? _____
4. Have the test limits been accurately identified? _____
5. Have the failure criteria been interpreted into meaningful test procedures? _____
6. Is there a specific EMI test procedure and detail including the operation of each monitoring and EMC test instrument for each test procedure? _____

General Requirements
1. Does the test plan acknowledge that the test report will be prepared? _____
2. Is there a sample test data sheet, including provisions for all critical system measurement equipment parameters, antenna correction factors, cable loses, etc.? _____
3. Are all important test arrangement diagrams, electrical loads, test equipment lists, etc., included? _____
4. Does the documentation methods proposed conform to MIL-STD-831 or the applicable documentation standard? _____

Source: *Electromagnetic Compatibility Design Guide for Avionics and Related Ground Support Equipment.* NAVAIR AD 1115. Washington, D.C.: U.S. Naval Air Systems Command, Department of the Navy, 1975.

Summary

- EMC is the ability of a system or product to function properly in its electromagnetic environment and not be a source of pollution. Susceptibility refers to the response of a system to interference from unwanted energy. Emission refers to the interference from a product or environment.
- EMC affects:
  ~ Product effectiveness.
  ~ Safety.
  ~ Regulatory compliance.
- Objectives for the EMC effort can be:
  ~ Conformance to performance requirements under all specified EMC environments.
  ~ Compliance with all regulatory requirements.
  ~ Minimize the cost of conformance by preferring prevention before correction.

- Four general classifications for EMC are:
  - ~ Radiated emissions.
  - ~ Conducted emissions.
  - ~ Radiated susceptibility.
  - ~ Conducted susceptibility.
- A special case of EMC is electrostatic discharge (ESD). The source is from static electricity.
- The main EMC program activities are:
  - ~ Setting the requirements.
  - ~ Preparing the EMC Control Plan.
  - ~ Preparing the EMC Test Plan.
  - ~ Providing analysis and solving problems.
  - ~ Designing for EMC.
  - ~ Verification of EMC.

Further reading

Dangelmayer, G. Theodore. *ESD Program Management*. New York: Van Nostrand Reinhold, 1990.

Heirman, Donald N., Robert C. Morris, and Steven M. Crosby. "Design for Multinational EMC Compliance." *AT&T Technical Journal*, Vol. 69, No. 3, May/June 1990, pp. 28-45.

Mardiguian, Michel. *Electrostatic Discharge*. Gainesville, Virginia: Interference Control Technologies, 1986.

Ott, Henry W. *Noise Reduction Techniques in Electronic Systems*. 2nd Ed. New York: John Wiley & Sons, 1988.

U.S. Naval Air Systems Command. *Electromagnetic Compatibility Design Guide for Avionics and Related Ground Support Equipment*. NAVAIR AD 1115. Washington D.C.: U.S. Naval Air Systems Command, Department of the Navy, 1975.

Welsher, Terry L., Timothy J. Blondin, G. Theodore Dangelmayer, and Yehuda Smooha. "Design for Electrostatic Discharge (ESD) Protection in Telecommunications Products." *AT&T Technical Journal*, Vol. 69, No. 3, May/June 1990, pp. 77-96.

11

Testability

Testability is important for building and maintaining products. Engineering for testability affects product costs, time to market, and product effectiveness. In this chapter, you'll learn:

- How to set testability requirements.
- What tasks the testability specialist performs.
- What should be in the testability program plan.

Defining testability

As the systems engineer, you should understand trade-offs in testability. Testability specialists perform activities for your project that must be managed. But what is testability? What does built-in-test mean? MIL-STD-1309C, *Definitions of Terms for Test, Measurement and Diagnostic Equipment*, has these definitions:

> Testability: A design characteristic which allows the status of a unit to be confidently determined in a timely fashion.
>
> Built-in-test: An integral capability of the mission equipment which provides an on-board, automated test capability to detect, diagnose, or isolate system failures. The fault detection and, possibly, isolation capability is used for periodic or continuous monitoring of a system's operational health, and for observation and, possibly, diagnosis as a prelude to maintenance action.

Testability allows quick and comprehensive verification of system functions during manufacturing and maintenance and before critical system operation.

Testability program activities

Testability program objectives include setting requirements, designing for testability, and verifying testability.

Setting requirements before design is essential. Testability cannot be effectively redesigned into a system that already exists. Testability must be designed into a system as an integral characteristic.

Design issues for testability include physical partitioning, functional partitioning, controllability, and observability. Partitioning allows interfaces and functions to be selected for efficient fault isolation. Controllability over internal items and devices is needed for detecting and isolating internal faults. Initialization to known states and resets are control measures. Observability includes test points, data paths, and nondisturbance of the system by the measuring devices. Built-in-test design trade-offs include:

- Centralized versus distributed.
- Tailored versus flexible.
- Active versus passive.
- Hardware versus software.

Verification includes predictions, analysis, and, finally, demonstration of testability effectiveness.

Benefiting from testability

Testability benefits include reduced test cost in the following areas:

- Production.
- Equipment.
- Programming.
- Test fixtures.
- Field maintenance.

Test effectiveness means:

- Reduced test time.
- Elimination of test bottlenecks.
- Reduced misses of failures.
- Reduced false alarms.
- Less training required.

Businesses survive by reducing costs and shortening the time to market. Long test times do not support these goals. Some causes of long test times are:

- Lack of product testability.
- Lack of test equipment.
- Lack of programming for the test equipment.
- Improper tests.
- Poor operator training.

Setting requirements

Setting requirements for testability is done in coordination with the manufacturing plan and the maintenance concept. In addition, usage and other environments must be defined. Quantitative and qualitative requirements are set from objectives, constraints, technology, and possible system configurations. Some major objectives may be to:

- Test effectively with minimum effort and cost.
- Reduce maintenance-induced problems.
- Speed time to market.
- Reduce the cost of test equipment and programming.
- Reduce cost of documentation.
- Prevent manufacturing bottlenecks.

Issues you should address as the systems engineer might include:

- Is the mission availability requirement satisfied?
- Are the testability requirements compatible with logistics support, maintenance concept, failure modes, and the manufacturing concept?
- Are the testability requirements consistent with the system constraints and project funding? Constraints can include:
  ~ Mobility requirements.
  ~ Warm-up time requirements.
  ~ Environments.
  ~ Availability of test equipment.
  ~ Ability of your designers.
  ~ Customer-preferred method of maintenance.

Quantitative requirements are derived and allocated from mission requirements, customer requirements, and system-level constraints. Typical testability values are shown in TABLE 11-1.

Table 11-1. Typical Testability Values.

	% Capability	Repair Level
Fault detection (all means)	90 – 100 100 100	Organizational Intermediate Depot
Fault detection (BIT)	90 – 98 95 – 98 95 – 100	Organizational Intermediate Depot

Fault isolation
  Eight or less modules
  Three or less modules
  One module
	
	95 – 100
	90 – 95
	80 – 90
		
		All
		All
		All

False alarms	1000 – 5000 hours between alarm	

Source: Systems Reliability and Engineering Division, Rome Air Development Center. *RADC Reliability Engineer's Toolkit.* Griffiss Air Force Base, NY: Systems Reliability and Engineering Division, Rome Air Development Center, Air Force Systems Command, 1988.

Specialty engineering

Qualitative requirements include banning the use of one-shots, requiring grounds assigned to certain connector pins, and requiring that feedback loops can be broken by the tester. These requirements can be imposed on the designers through guidelines and handbooks.

Tasking the testability specialist

MIL-STD-2165, *Testability Program for Electronic Systems and Equipments*, offers guidance on conducting a testability program. The tasks are to be tailored for the particular needs of the system. The goals of the testability specialist, as described by MIL-STD-2165, are:

a. Preparation of a Testability Program Plan.
b. Establishment of sufficient, achievable, and affordable testability, built-in and off-line test requirements.
c. Integration of testability into equipments and systems during the design process in coordination with maintainability design process.
d. Evaluation of the extent to which the design meets testability requirements.
e. Inclusion of testability in the program review process.

Tasks in MIL-STD-2165 are grouped into three sections:

1. Task Section 100—program monitoring and control
2. Task Section 200—design and analysis
3. Task Section 300—test and evaluation

These tasks are summarized for you in this section. The purpose of each task is taken directly from MIL-STD-2165. You must judge whether the testability specialist is meeting the top-level requirements.

Testability program planning (task 101)

The purpose of task 101 is to plan for a testability program that identifies and integrates all testability design management tasks required to accomplish program requirements. A description of the subtasks is under the "Controlling testability" heading of this chapter.

Testability reviews (task 102)

The purpose of this task is to establish a requirement for the performing activity to: (1) provide for all official review of testability design information in a timely and controlled manner, and (2) conduct in-process testability design reviews at specified dates to ensure that the program is proceeding in accordance with the contract requirements and program plans.

This task asks for testability to be included in each system review. Reviews of testability should include the:

- Status and results of testability tasks.
- Testability-related requirements.
- Testability design, cost, or schedule problems.

Design reviews must cover the:

- Impact of the diagnostic concept on readiness, life cycle costs, personnel, and training.
- Built-in-test.
- Rationale for testability criteria and weighting factors.
- Testability techniques used by the design engineers.
- Failure Modes and Effects Analysis as a basis for test design.
- Coordination between hardware and software efforts for test.
- Compatibility of test signals with test equipment.
- Approaches to production and field testing.

Testability data collection and analysis planning (task 103)

The purpose of task 103 is to establish a method of identifying and tracking testability-related problems during system production and deployment and identifying corrective actions. The three major subtasks are:

1. *Develop a plan for analysis of production test results* to determine if testability requirements are being met.
2. *Develop a plan for analysis of field maintenance* to determine if testability requirements are being met.
3. *Define data collection requirements* to meet the needs of the testability analysis.

Testability requirements (task 201)

The purpose of task 201 is to recommend system test and testability requirements that best achieve availability and supportability requirements and to allocate those requirements to subsystems and items. This task requires inputs from task 203 of MIL-STD-785 and task 205 of MIL-STD-470 for reliability and maintainability analyses.

The task establishes top-level testability design objectives, goals, and constraints. It also identifies technology, existing and planned test resources, and problems on similar systems. It provides for requirements flow-down and allocation. Trade-off studies and risk estimation are done under this task. Examples of trade-offs are:

- Sensitivity of readiness parameters to variation in testability parameters.
- Sensitivity of life cycle costs to variation in testability parameters.

- Alternative diagnostic concepts effects on maintenance hours, skill levels, and experience.
- System downtime costs because of erroneous failure indications.

Testability preliminary design and analysis (task 202)

The purpose of task 202 is to incorporate testability design practices into the design of a system or equipment early in the design phase and to assess the extent to which testability is incorporated. This should be done in the concept definition stage.

Task 202 analyzes and evaluates alternative design concepts to ensure testability is supported. The analysis identifies the presence or absence of features that facilitate testing. Modification or selection of alternative design concepts is less expensive earlier in the project life cycle. Finally, a model for prediction of fault detection and fault isolation is developed.

Testability detail design and analysis (task 203)

The purpose of task 203 is to incorporate features into the design of a system or equipment that satisfies testability performance requirements and predicts the level of test effectiveness achieved.

The testability specialist assists design engineers as they incorporate testability into the design. The testability specialist analyzes the designs using failure modes and predictions. The calculated test effectiveness is compared to the testability requirements, and the design is iterated until all requirements have been met. It is extremely important that testability is designed into the system and not retrofitted or patched in later.

Cost data is assembled for system design, test-generation efforts, and production test. This data, as well as test effectiveness predictions, are fed into life cycle cost estimates.

Testability inputs to maintainability demonstration (task 301)

The purpose of task 301 is to determine whether specified testability requirements have been complied with and to assess the validity of testability predictions. A Testability Demonstration Plan is the main output of this task. It should address how to verify the:

- Ability of operational system checks to detect failures.
- Ability of system built-in-test to detect and isolate failures.
- Compatibility of each unit with the test equipment.
- Ability of test equipment to detect and isolate failures.
- Adequacy of documentation.
- Validity of models used to predict testability parameters.

MIL-STD-2165 summary

MIL-STD-2165 tasks are starting points for a testability program. To be effective, however, the tasks must be tailored to the particular application. A qualified testability specialist is needed to manage the testability effort. Analysis, standardization, and coordination of the testability program results in cost and time savings.

Guiding the design

Too many times, the need for testability is not discovered until production has already started. Design engineers are essential to the success of testable products, and fortunately, the body of knowledge is growing rapidly in this field. Design engineers need training, access to testability specialists, guidelines, and checklists so that they can meet requirements. System engineering and management must expect and require that testability be part of the product requirements.

Tasking design engineers

You should ensure that design engineers are meeting requirements for testability. Questions you should be asking include:

- Do the design engineers have design guidelines and checklists and know how to apply them?
- Do the design engineers understand the test, manufacturing, and maintenance philosophy; and the intended test equipment?
- Are the designs documented for testability predictions and calculations?
- Have all design engineers been given quantitative- and qualitative-allocated requirements for testability? Do they understand them, and are they meeting them?
- Do individual designers present testability as part of their design reviews, showing predicted testability levels?

A sample partial checklist for electronic equipment designers is shown in TABLE 11-2.

Table 11-2. Partial Testability Checklist.

Mechanical Design	
Are all components oriented in the same direction?	_____
Are standard connector pin positions used for power, ground, clock, etc.?	_____
Are connector pins assigned so that any shorting of physically adjacent pins will cause minimum damage?	_____
Is the design free of special setup (e.g., special cooling), which would slow testing?	_____
Is each hardware component clearly labeled?	_____
Partitioning	
Is each function to be tested wholly upon one board?	_____
If more than one function is on one board, can each be tested independently?	_____

Table 11-2. Continued.

Within a function, can digital and analog circuitry be tested independently? _____

Test Control

Can circuitry be quickly and easily driven to a known initial state? _____

Is it possible to disable on-board oscillators and drive all logic using a tester clock? _____

Can the tester electrically partition the item into smaller, independent, easy-to-test segments? _____

Is circuitry provided to bypass one-shot circuitry? _____

Can feedback loops be broken under control of the tester? _____

In microprocessor-based systems, does the tester have access to the data bus, address bus, and important control lines? _____

Test Access

Are signal lines and test points unaffected by the capacitive loading of the test equipment? _____

Are buffers or divider circuits employed to protect test points that could be damaged by an inadvertent short circuit? _____

Are high voltages scaled down within the item prior to providing test point access to be consistent with tester capabilities and human safety? _____

Is the measurement accuracy of the test equipment adequate compared to the tolerance requirement of the item being tested? _____

Parts Selection

Is the number of different part types the minimum possible? _____

Have parts been selected that are well characterized in terms of failure modes? _____

Is a single logic family being used? _____

Adapted from: U.S. Department of Defense. *Testability Program for Electronic Systems and Equipments.* Military Standard 2165. Washington, D.C.: U.S. Department of Defense.

Controlling testability

The Testability Program Plan provides a systematic approach to testability. Systems engineers work with the testability specialist to set top-level requirements. The plan should be prepared by the testability specialists, with approval by systems engineering. The plan is used to:

- Manage the testability program.
- Evaluate the project's understanding and execution of testability.
- Evaluate planning for analysis and coordination of data distribution.
- Evaluate the staffing and organization of the testability specialists.

MIL-STD-2165 describes the purpose of the Testability Program Plan as ". . . To plan for a testability program which will identify and integrate all testability design management tasks required to accomplish program requirements."

The remaining sections—"Task description," "Task input," and "Task output"—are all taken from MIL-STD-2165.

Task description

The first step is to identify a single organizational element within the performing activity that has the overall responsibility and authority for implementing of the testability program. Establish analyses and data interfaces between the organizational element responsible for testability and other related elements.

Develop a process by which testability requirements are integrated with other design requirements and disseminated to design personnel and subcontractors. Establish controls for ensuring that each subcontractor's testability practices are consistent with overall system or equipment requirements.

Identify testability design guides and testability analysis models and procedures to be imposed on the design process. Plan for the review, verification, and utilization of testability data submissions.

Finally, develop a testability program plan that describes how the testability program is to be conducted. The testability program plan must be included as part of the Systems Engineering Management Plan or other integrated planning documents when required. The plan describes the time phasing of each testability task included in the contractual requirements and its relationship to other tasks.

Task input

Task inputs require identifying each testability task required as part of the testability program; identifying the time period each task is to be conducted; identifying approval procedures for plan updates; and identifying deliverable data items.

Task output

The Testability Program Plan, in accordance with DI-T-7198, if specified, is a stand-alone plan. When required to be a part of another engineering or management plan, use the appropriate, specified Data Item Description.

Summary

- Testability is a design characteristic that allows the status of a unit to be confidently determined in a timely fashion.
- Testability allows quick and comprehensive verification of system functions during manufacturing and maintenance and before critical system operation.
- Design issues for testability include physical partitioning, functional partition-

156 **Specialty engineering**

ing, controllability, and observability. Testability benefits include reduced test costs for:
- Production
- Equipment
- Programming
- Test fixtures
- Field maintenance
- Some major testability objectives might be to:
  - Test effectively with minimum effort and cost.
  - Reduce maintenance-induced problems.
  - Speed time to market.
  - Reduce cost of test equipment and programming.
  - Reduce cost of documentation.
  - Prevent manufacturing bottlenecks.
- MIL-STD-2165, *Testability Program for Electronic Systems and Equipments*, offers guidance on conducting a testability program. Task 101 of MIL-STD-2165 describes the Testability Program Plan.

Further reading

U.S. Department of Defense. *Testability Program for Electronic Systems and Equipments*. Military Standard 2165. Washington, D.C.: U.S. Department of Defense.

12

Software

Software now accounts for a large portion of many systems. Therefore, the software effort must follow the systems engineering process. Software is important because it affects risk, schedule, cost, performance, maintainability, and reliability. In this chapter, you'll learn:

- How to specify requirements for software.
- Plan for software.
- Control the software effort.

This chapter does not explain the software process, accepted design practice, or design. The books listed in "Further reading" can help you manage software. Many new references are appearing every week on software processes and design, and it is impossible to explain software design in one chapter.

Requirements

Poorly defined requirements are a sure recipe for disaster when software is involved. This section describes some characteristics of software requirements. Reliability and maintainability are defined specifically for software.

Software requirements characteristics

As described in IEEE Std 830-1984, characteristics of good software requirements are:

- Unambiguous
- Complete
- Verifiable
- Consistent
- Modifiable
- Traceable

Unambiguous means that every requirement has only one interpretation. Requirements written in English are susceptible to ambiguity due to word order and emphasis. Therefore formal requirements languages can be used to reduce ambiguity.

Complete means including all significant requirements, functions, behaviors, performance, constraints, and interfaces. Valid and invalid input values have specified responses.

Requirements are verifiable when a cost-effective means exists for people or machines to check that the product meets the requirement. Approximately, usually, resistant, and fail-soft are terms that are *not* verifiable.

Consistent requirements are not in conflict with each other. Different terms, different characteristics, and different sequences are examples of conflicts between specification requirements.

Modifiable requirements are easy to change completely and consistently. Modifiable is assisted by an index and cross-referencing, and only stating a requirement once.

Traceable means that the origin of the requirement is clear. Upward traceability means that the requirement can be traced through allocation or derivation back to customer needs. Downward traceability finds the impact to lower-level requirements caused by top-level changes.

Expressing requirements

Software requirements should specify the needs and not the solutions. Basic issues include:

- Functionality—what it is supposed to do.
- Performance—bits processed, speed, response time, etc..
- Constraints—standards, data format, languages, etc..
- Attributes—portability, maintainability, correctness, etc..
- Interfaces—software, hardware, and people.

Inputs, outputs, and transformations are included in requirements. Inputs include sources, timing, valid ranges, and format. Transformations include validity checks, timing, sequence of operations, and response to abnormal situations. Outputs include destinations, timing, formats, and error messages.

There are many attributes used to describe software. TABLE 12-1 lists some commonly found attributes. The environment affects attribute importance. For example, human safety requires that the emphasis be on the attributes of reliability, correctness, and verifiability.

Requirements flow-down

Software requirements flow-down follows the systems engineering process. Some of the information inputs may appear different to you, however. The top-system definition still looks at user needs, interfaces, and logistics concepts. At the point of allocation to hardware and software, some differences begin. The person-machine interfaces are different in some ways. The throughput analysis and flow of data are software-tailored.

Table 12-1. Examples of Software Attributes

Functionality	Capabilities
	Completeness
	Verifiability
Useability	Ease of use
	Consistency
	Ease of training
	Input preparation
	Output interpretation
Reliability	Frequency of failure
	Severity of failure
	Recovery from failure
Performance	Speed
	Response time
	Efficiency
Supportability	Ease of upgrade
	Ease of repair
	Installability
	Expandability
	Testability
	Flexibility
	Portability

The really big break occurs when the software's top-level design document is written. The hardware interface is seen from a software view. The architecture, timing, storage, and recovery definitions are software flavored. As detailed design documents are made, there are software-to-software interfaces, performance budgets, data dictionaries, database designs, and other requirements that are different than hardware. The requirements flow-down process is the same; it is information that causes an apparent difference. Software must not be treated as a magical art form that cannot be subjected to a rigorous method of development.

Reliability

ANSI/IEEE Std 729-1983 defines software reliability as "The probability that software will not cause the failure of a system for a specified time under specified conditions." You can easily confuse reliability and availability when dealing with software. A failure might happen briefly and then be swept away by the torrent of computer instructions. The system might be available because the failure did not cause a crashing halt, nevertheless, a failure occurred. Any situation where a program does not function to meet user needs is a failure.

You may find arguments that software doesn't wear out or break and that you only get errors, not failures. This misses the point of view of the customer. The customer doesn't care if it is an error or "little bug" in the product. If the product does not function correctly, it is a failure to the customer.

Maintainability

Software maintainability is a measure of how quickly a system with a fault can be brought to a reliable state. This includes identifying the fault, repairing the fault, recovering from the effects of the fault, and verifying that all requirements are still being met.

Maintainability can be helped by standardizing of notations, modularity, use of comments, and interface documentation. Using "tricks" and undocumented shortcuts might speed execution but maintainability suffers. Test tabs and flags ease development and later modifications. Code clarity and standards for indentation, parameters, and error-handling save valuable troubleshooting time.

Technical planning

You must expect and demand that the software specialists have:

- Competent people.
- Problem-solving behaviors.
- A process for software development.
- Quality assurance of that process.

Without these basics, planning is extremely difficult because no one knows what to do. Finding competent people should be your first priority. Software production is highly dependent on staff experience, requirements completeness, interface complexity, and stability of the requirements and verification. Experienced staff have been found to be twice as productive as inexperienced staff. Planning must be done by experienced people and not assigned to someone new.

Software tasks

The software specialist plans the software tasks. Detailed definition of tasks is necessary for control and early warning of trouble. TABLE 12-2 lists some of the typical software tasks that appear in Work Breakdown Structures.

Estimating

For most companies, software estimates are notoriously understated. Often, the time for testing and documentation has been ignored. Frederick Brooks, author of *The Mythical Man-Month*, suggests this rule of thumb for scheduling software development time:

1/3	planning
1/6	coding
1/4	component/early system test
1/4	system test

Typically, both software and hardware developers estimate optimistic times for testing during the component and integration phases. When looking at costs, an indus-

Table 12-2. Software Tasks.

- Define software objectives.
- Define software constraints.
- Define project standards.
- Plan development phases.
- Define development controls.
- Define resource requirements.
- Staff the project.
- Write the Software Development Plan.
- Plan reviews, walk-throughs, and audits.
- Develop the software work breakdown structure and dictionary.
- Select software tools.
- Select languages and support.
- Define training needs for staff.
- Define schedules to meet project deadlines.
- Define software interfaces.
- Define configuration control.
- Assign work packages.
- Identify end product formats.
- Tailor methodologies.
- Verify the product against the requirements.

try rule of thumb is to multiply the software staff's estimate by three. Unless a formal estimating process is in place, software people tend to estimate only low-level design and coding. They typically leave out:

- Systems engineering support.
- Top-level software requirements definition.
- Requirements allocation and flow down.
- Reviews and walk-throughs.
- Software integration and test.
- Hardware engineering support.
- Systems integration and test.
- Supervisor time.
- Configuration management.
- Quality assurance.
- Documentation and publications.

TABLE 12-3 shows some of the factors affecting software costs. Some guidelines to estimating are:

- Never ask an inexperienced person to estimate.
- Use teams to estimate, if possible.

- Understand your risk.
- More detailed usually means more accurate.
- Write the assumptions, and what is *not* included.
- Test the estimates for reasonableness.

Table 12-3. Factors Affecting Software Costs.

- Developers' experience.
- Stability and completeness of requirements.
- Language requirements.
- Interfaces.
- Schedule.
- Processor time constraints.
- Design complexity.

Technical control

The head software specialist is responsible for the software effort, but because you are responsible for the system, you should watch for:

- *The right product.* Software is being built to the requirements.
- *The right process.* Software is being built using standards and a software development process.
- *The schedule.* Progress is tracking the work planned.
- *People.* People are happy, they are not burning out, they are communicating, and productivity is good.

Disciplined design

Sooner or later, you'll run into a story that goes much like this: "You can't expect our (gurus, artists, magicians) to follow all this structure (MIL-STD-2167, quality assurance plan, work planning). They must have freedom to be creative!" Granted, there is a lot of creativity needed to produce a system or product. There also must be discipline in the creation, however. Otherwise, you are not running a business but a hobby. Enforce your standards.

Inspections

One technical control technique that has a high payoff for software is inspections. Inspections and walk-throughs are done many times in the development process. They catch errors earlier, while they are less expensive to fix. Inspections should be done by peers and not by supervisors. The inspection meeting should be less than two hours. The meeting leader must understand the inspection technique and, preferably, should be trained. The results of the inspection are documented. Remember, inspections are problem-identification meetings, not problem-solving sessions.

Verifying that requirements are met

"How many lines of code have you done?" One parameter to be wary of is lines of code. It is easy to measure but really tells you nothing about meeting the requirements. If you were told that the system had x transistors or y bolts, would you know if the requirements were being met? No, and you can't do any better by knowing lines of code. Measure verified modules against requirements and functionality to have a better picture of progress.

Managing risk

Warning signs of risk in software development are:

- Changing requirements.
- Inexperienced and insufficient staff.
- Aggressive schedules that are calendar-driven.
- Lack of communication.
- Poor management control system.

To counteract changing requirements, follow the systems engineering process. Don't allow coding until requirements have been determined. Force hardware and software designers to talk with each other.

To do the job, you must have the proper people. Inexperience in software or the application will show in the work. Use work breakdown structures, resource identification, and leveling to plan for the people needed. It would be nice if schedules were driven by events and not by calendar dates. Critical paths must be identified in the schedules of work to be done. While aggressive schedules can be a fact of life, risk management can prevent some problems from occurring.

Communication problems can exist inside the software group, with hardware specialists, and with the external world. Co-locating the members of the design team is essential. A team atmosphere must be created so that "us" and "them" do not exist. Face to face talking will solve many of the communication problems.

Poor management control systems have a symptom. The symptom is that you don't know and can't determine where the project stands in achieving its objectives. After several weeks or months of reporting that everything is 90 percent complete you slowly realize you don't know what is happening. Part III describes managing the systems engineering process. Work breakdown structures, resource identification, resource leveling, critical paths, and other tools needed to keep the project heading towards completion are discussed.

Delegate

It is difficult to control software development if you are looking after the system engineering. There is not enough time to do both. Unless you have a software background, you also don't know what to watch out for. Find a competent software specialist who can manage the software effort. Structure the controls such that progress or lack of progress becomes rapidly evident. Continue asking questions of the people who are doing the work.

Summary

- Software is important because it affects:
  - Risk
  - Schedule
  - Cost
  - Performance
  - Maintainability
  - Reliability
- Poorly defined requirements are a sure recipe for disaster when software is involved.
- Software specialists should have:
  - Competent people.
  - Problem-solving behaviors.
  - A process for software development.
  - Quality assurance of that process.
- Experienced staff have been found to be twice as productive as inexperienced staff.
- When looking at costs, an industry rule of thumb is to multiply the software staff's estimate by three.
- Warning signs of risk in software development are:
  - Changing requirements.
  - Inexperienced and insufficient staff.
  - Aggressive schedules that are calendar-driven.
  - Lack of communication.
  - Poor management control system.
- You are running a business, not a hobby. Enforce your standards.
- Inspections should be done by peers and not by supervisors.

Further reading

Brooks, Frederick P., Jr. *The Mythical Man-Month: Essays on Software Engineering*. Reading, Massachusetts: Addison-Wesley Publishing Company, 1975.

Evans, Michael W. and John J. Marciniak. *Software Quality Assurance and Management*. New York: John Wiley and Sons, 1987.

Gilb, Tom. *Principles of Software Engineering Management*. Wokingham, England: Addison-Wesley Publishing Company, 1988.

Humphrey, Watts S. *Managing the Software Process*. Reading, Massachusetts: Addison-Wesley Publishing Company, 1989.

Rakos, John J. *Software Project Management for Small to Medium Sized Projects*. Englewood Cliffs, New Jersey: Prentice Hall, 1990.

13

Producibility and manufacturability

Producibility and manufacturability are essential ingredients in transforming engineering drawings into physical products. In this chapter, you'll learn:

- Producibility basics.
- How producibility benefits your project.
- Key elements of a producibility program.

Producibility is important to minimize:

- Component and assembly costs.
- Development cycle times.
- Assembly errors and manufacturing-induced defects.

Defining producibility

Producibility enables manufacturing to fabricate your product at a minimum cost without sacrificing quality. Producibility is a measure of the relative ease of producing a product. A producible design includes engineering and manufacturing consideration of:

- Material selection.
- Tooling.
- Facilities.
- Capital equipment.
- Test equipment.
- Methods.
- Processes.
- People.
- Production quantities.
- Production rates.

166 Specialty engineering

Producibility objectives include:

- Maximizing:
  - Simplicity of design.
  - Ease of assembly and integration.
  - Process repeatability.
  - Robustness against manufacturing noise.
  - Simplicity of processes.
  - Use of economical materials.
  - Use of economical manufacturing technology.
  - Standardization of materials and components.
  - Product inspectability.
- Minimizing:
  - Cycle times.
  - Manufacturing bottlenecks.
  - Unit costs.
  - Variation about target values.
  - Scrap and rework.
  - Number and variety of parts and materials.
  - Energy consumption.
  - Special testing.
  - Use of critical processes.
  - Skill levels of production people.
  - Design changes.

Design for manufacturing

In a broader definition, design for manufacture (DFM) is concerned with all interactions of the manufacturing system. Knowledge of the interactions is used to optimize the manufacturing system for quality, cost, and delivery. Part of this is understanding how a product's physical design interacts with the manufacturing system to facilitate global optimization of the product and the manufacturing system together. The objectives of design for manufacture are:

- Identify product concepts that are easy to manufacture.
- Design components for ease of manufacture and assembly.
- Integrate the product design and the manufacturing system design to match needs and requirements.

Figure 13-1 shows a simplified interface between engineering and manufacturing. The end products of engineering are drawings. These drawings contain requirements that are passed on to manufacturing. Manufacturing deals with knowledge, materials, labor, and energy to convert requirements into finished products. Processes and assembly are used in the transformation.

Design engineering must seek robustness against the manufacturing noise. This could be poor quality parts or material. It could also be the worker who is feeling ill or

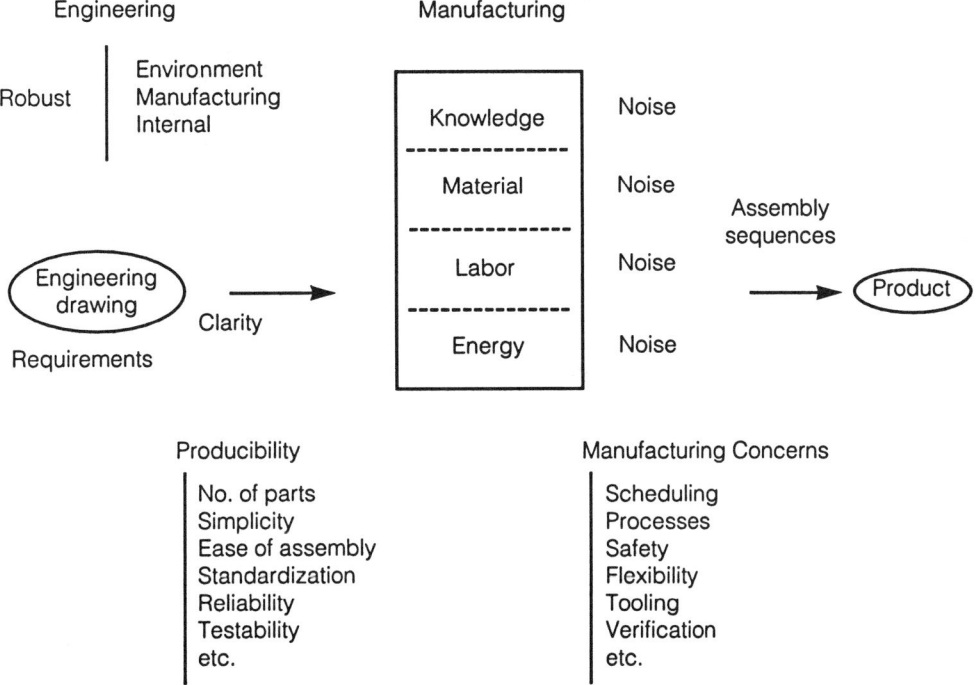

Fig. 13-1. Simplified engineering/manufacturing interface.

worried about home life. Design engineering increases the latitude for the manufacturing processes to have greater success.

Manufacturing, on the other hand, seeks to decrease variance about design goals. The proper suppliers must be selected. The processes must be capable of holding about the design points. Workers must be trained to understand what affects quality. Manufacturing must be flexible to overcome noise in the process.

But what is flexible? You could say it means the ability to handle engineering changes, process changes, or customer order changes. But we are really asking, can manufacturing build to our order? The only way to do that consistently is to have a predictable and rapid build-time. And to do this means to have rapid throughput with small wait-times. Design engineering must work with manufacturing so the product will achieve rapid throughput in the manufacturing environment.

It is clear that the systems engineering process must include producibility as criteria. Looking at FIG. 13-2, you can see that manufacturing concerns can appear as criteria or constraints, either implicitly or explicitly, in the synthesis process.

Benefiting from producibility

Most companies have two business goals: (1) make a good product, for the least cost, as quick as you can; and (2) to sell as many as you can, as fast as you can, for as much as

168   Specialty engineering

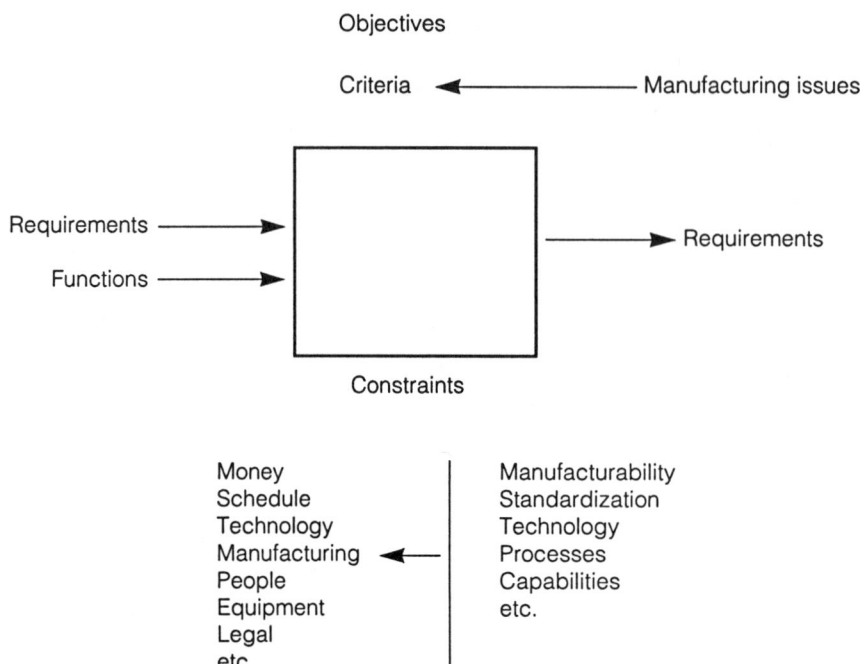

Fig. 13-2. Manufacturing issues affect criteria and constraints.

possible. Producibility supports these goals. Producibility seeks to:

- Minimize costs.
- Minimize development cycles.
- Maximize robustness in the manufacturing process.
- Minimize assembly errors, which also affects:
  ~ Product defects
  ~ Reliability
  ~ Safety
  ~ Warranties

Implementing producibility

Implementing producibility means planning for producibility. Systems engineering is really decision-making, and planning is making decisions. In your plan, decide on the objectives, the specialists needed, and the time phasing of the efforts. Producibility should start at the systems level, continue through subsystems, and increase during the component design phase. Set up a communication and documentation system among design engineering, producibility, and manufacturing. Training for design engineers in producibility principles is necessary, along with checklists and other job aids. Manufacturing capabilities should be explicitly stated. Ideally, a producibility specialist will be on hand to assist design engineers during all phases of the system design.

Producibility tasks

The producibility team's tasks should include:

- Ensuring that manufacturing technology needed is available or adequately planned for.
- Analyzing and updating risk assessments and alleviations.
- Continually assessing the design for producibility considerations.
- Standardizing parts and materials.
- Performing trade studies.
- Evaluating plans for proofing production processes, tooling, and test equipment.
- Producing and updating the producibility plan.

Several approaches to manufacturing integration have been recommended by practitioners. These include:

- Having manufacturing engineering on the project from the first day.
- Providing manufacturing feedback on proposed designs before fabrication.
- Providing design guidelines and training.
- Using Taguchi and Design of Experiments techniques.
- Rating designs using quantitative methods such as Boothroyd-Dewhurst and Hitachi.

Design engineers are sometimes overcome by the technical difficulty of solving immediate problems and forget about the later fabrication of their design. There are critical manufacturing decisions just as there are critical technical decisions. Manufacturing requires creating an assembly sequence, identifying subassemblies, deciding when to test and inspect, and designing parts so that functional and tooling tolerances are compatible with the method of assembly.

Design for assembly

Parts cannot be optimized in isolation. Design for Assembly (DFA) considers the integration of parts into assemblies and systems. Quantitative techniques are used to score ease of assembly. The most well-known is Boothroyd-Dewhurst's software for mechanical assemblies and printed circuit boards. It has been claimed that DFA has reduced printer assembly time from thirty minutes to three minutes. The approach analyzes each part for:

- Necessity as a separate part.
- Ease of handling, feeding, and orienting.
- Ease of assembly.

Results include an estimate of the assembly time and a rating for design efficiency.

The other technique used by major corporations is the Hitachi/GE methodology. The "Assemblability Evaluation Method" was developed in Japan after one of Hitachi's engineers attended a Boothroyd seminar. General Electric has made the methodology available to selected U.S. companies under license from Hitachi.

Best design practice

In general, the best design practice includes:

- Simplicity of design—cost is proportional to number of parts
- Standardization of parts and materials—fewer part numbers mean better control over suppliers and internal handling.
- Manufacturing process capability analysis—process capability must be compared to tolerance specification.
- Design flexibility—designs should offer the maximum number of alternative materials and processes to produce the item.

Design affects cost

Designers must understand the effect of product design on costs. From his work in Hewlett-Packard's Manufacturing Research Center, Douglas Daetz (*Quality Progress*, June 1987) made the following observations:

- Assembly time (cost) is proportional to the number of parts assembled.
- Material costs can usually be reduced by reducing the number of parts.
- The system cost of carrying a part number in a manufacturing division might range from $500 to $2,500 annually, not counting the cost of the parts themselves.
- Design changes, on average, after release to manufacturing incur a system cost of $5,000 to $10,000 per design change.

Good design practice

Good design practice will result in:

- Minimizing the number of different parts.
- Minimizing the total number of parts.
- Minimizing the number of suppliers for procured parts.
- Complete and accurate specifications to the supplier.
- Use of preferred parts.
- Designs that are relatively free from tolerance-dependence.
- Self-aligning parts and mistake-proof assemblies.
- Parts robust to handling, contamination, corrosion, and cleaning.
- Adjustment-free assemblies.

Process capability analysis

Analyze the manufacturing processes for:

- Compatibility of the raw materials with the process.
- Shape and size restrictions for the process.
- Production rate of the process.
- Process tolerance capability.

- Tooling requirements.
- Labor.
- Waste rates and rework.
- Optimum lot sizes.

Assessing risk

Risk is usually highest when you have a new product or a new process. Assess the risk of manufacturing processes. The process will be state of the practice, state of the art, or experimental. State of the practice is well understood and has a long usage record. Risk is generally low. State of the art is somewhat new and is available from a limited number of sources. Risk is higher. Experimental means the process has been demonstrated in the laboratory but not in the real factory. Risk is highest.

Design checklists

Producibility specialists should prepare design checklists specific to your project. An example of general questions is shown in TABLE 13-1. The list is incomplete. A qualified producibility specialist must tailor your own guidelines and checklists. No checklist will replace an innovative, knowledgeable engineer.

Table 13-1. Partial General Design Producibility Checklist.

1. Have alternative design concepts been considered and the simplest and most producible one selected? _____
2. Does the design exceed the manufacturing state-of-the-art? _____
3. Is the design conducive to the application of economic processing? _____
4. Does a design already exist for the item? _____
5. Is the item overdesigned or underdesigned? _____
6. Can redesign eliminate anything? _____
7. Is motion or power wasted? _____
8. Can the design be simplified? _____
9. Can a simpler manufacturing process be used? _____
10. Can parts with slight differences be made identical? _____
11. Can compromises and trade-offs be used to a greater degree? _____
12. Is there a less costly part that will perform the same function? _____
13. Can a part designed for other equipment be used? _____
14. Can weight be reduced? _____
15. Is there something similar to this design that costs less? _____
16. Can the design be made to secure additional functions? _____
17. Can multiple parts be combined into a single part? _____

Adapted from: *Design Guidance for Producibility*. Military Handbook 727. Washington, D.C.: U.S. Department of Defense.

Essential producibility

The most essential step toward effective producibility is having producibility and manufacturing on the design team from the start of the project. You must provide the means for them to evaluate the design and improve it for producibility. A meaningful way to communicate to design engineers is to construct a two-dimensional chart with manufacturing steps and product design characteristics. The characteristics can then be evaluated for positively or negatively influencing the manufacturing process. Manufacturing constraints in the systems engineering process cannot be ignored.

Summary

- Producibility is a measure of the relative ease of producing a product.
- Producibility is important to minimize:
    - Component and assembly costs.
    - Development cycle times.
    - Assembly errors and manufacturing-induced defects.
- The objectives of design for manufacture are:
    - Identifying product concepts that are easy to manufacture.
    - Designing components for ease of manufacture and assembly.
    - Integrating the product design and the manufacturing system design to match needs and requirements.
- In general, the best design practice includes:
    - Simplicity of design.
    - Standardization of parts and materials.
    - Manufacturing process capability analysis.
    - Design flexibility.
- The most essential step toward effective producibility is having producibility and manufacturing on the design team from the start of the project.

Further reading

Daetz, Douglas. "The Effect of Product Design on Product Quality and Product Cost." *Quality Progress*, June 1987, pp. 63-67.

U.S. Department of Defense. *Design Guidance for Producibility*. Military Handbook 727. Washington, D.C.: U.S. Department of Defense.

14
Value analysis

Value analysis is a system for identifying problems, then selecting and identifying problem-solving procedures. Value analysis is important because it seeks to minimize the cost of required functions and discourages single-point design solutions. Both producibility specialists and design-to-cost use value analysis. In this chapter, you'll learn the:

- Principles of value analysis.
- Benefits of value analysis.
- Steps in value analysis.

Defining value analysis

You might have seen or heard of the terms *value analysis*, *value engineering*, or *VA/VE*. Consider the terms as having the same meaning, and, for simplicity, this chapter uses the term *value analysis*. The objective of value analysis is to deliver to the customer (and user) the required functions at a minimum cost.

Value for engineering purposes is defined as:

$$\text{value} = \frac{\text{worth}}{\text{cost}}$$

You can measure this better if the equation is written:

$$\text{value} = \frac{\text{customer acceptance}}{\text{life cycle cost}}$$

Value analysis is credited to Lawrence Miles, a General Electric engineer. In the late 1940s, he developed the system and it was used at General Electric. The U.S. Department of Defense adapted the system, calling it *value engineering* and taught the principles through the U.S. Army Management Engineering Training Activity.

Benefiting from value analysis

Value analysis has two primary benefits:

1. It seeks to deliver functions at a minimum cost.
2. It shifts the viewpoint of the problem solver from single-point solution thinking to unconstrained solution thinking.

Implementing value analysis

The general approach to value analysis is to:

1. Break down the product into separate parts and functions.
2. Determine the cost for each part and function.
3. Determine the relative value of each part or function in the end product.
4. Search for a new approach for items out of balance in cost and value.

A step-by-step process to value analysis is:

1. Select the product, subsystem, or component for value analysis application.
2. Determine the functions and pull together information about the requirements, history, and present costs.
3. Cost the functions.
4. Determine the function's worth.
5. Identify value analysis targets by imbalance between cost and worth.
6. Create ideas, move into concepts, and refine concepts.
7. Analyze the alternative concepts for cost and requirements satisfaction.
8. Test and verify the alternative.

Balance means that low-worth items should not have high cost, for example.

Value analysis team

Five members is the ideal size for a value analysis team. The team might include a:

- Design engineer for product under study.
- Manufacturing engineer.
- Cost estimator.
- Producibility engineer.
- Marketer or salesperson.

The team must have:

- A complete set of engineering drawings.
- A real or model hardware.
- Indented and costed bill of materials and labor.

- A manufacturing process routing sheets.
- Access to competitive product information (internal and external).
- The history and knowledge of customer requirements.

Function analysis

Function analysis is fundamental to value analysis. By describing the functions of the product rather than the physical parts, thinking about solutions is less constrained. It is more desirable describing what the customer wants than describing what the item does. As you have seen in the systems engineering process, functions are really requirements. Therefore, it is critical to describe the correct requirements.

Functions are described by active verb and noun combinations. These are the rules:

- One active verb, one noun.
- Do not use "be" or "provide."
- The noun is not a part, activity, or operation.
- Have the viewpoint of the user.

It is easy to become confused if you allow the functions to become mixed. For example, mixing operator functions with the product functions. Functions can be analyzed for the product, the operator, the support equipment, and facilities. Do not mix these together in one analysis or you will be confused.

Teams use different techniques depending on their backgrounds. Brainstorming using formal rules helps uncover functions that could be overlooked. Teams often use Post-it note pads to sort and order functions. Functions are eventually organized in flow diagrams to show the relationships and sequences.

Functions can also support customer desires such as reliability, ease of use and maintenance, and aesthetics. These are very valid functions if desired by the customer and the user.

Value analysis checklist

The following value analysis checklist can help you ensure that the right questions have been asked before committing to a design. Some questions you might want to include:

1. Are all specified characteristics requirements?
2. Will a modification of the part simplify designer fabrication?
3. Can this function be eliminated?
4. What does it do unnecessarily?
5. Is it overdesigned (does it do more than required)?
6. Are there alternative manufacturing processes (is it designed for flexibility)?
7. Are there special requirements for installation, maintenance, testing, or safety?
8. Are costly or long lead time materials used?
9. Are the materials difficult to handle or process? Are they hazardous?
10. How is the part made? Why?

11. Who else makes something similar?
12. Are there any special problems with handling, packaging, storing, or transporting the item?
13. Can the design be changed to eliminate parts?
14. Can the design be purchased more cheaply?
15. Is there a less costly part that will perform the same function?
16. Can two or more parts be combined?

Using value analysis

If you have never performed value analysis before, you should choose a small item to begin the process. You would be wise using a competent facilitator until your teams learn the process. You might have decided that performing value analysis is easier on existing designs than on new designs. For new designs, the concept of design to cost is introduced in chapter 15.

Summary

- Value analysis is important because it:
  - ~ Seeks minimum cost for required functions.
  - ~ Discourages single-point design solutions.

$$\text{value} = \frac{\text{customer acceptance}}{\text{life cycle cost}}$$

Further reading

Fowler, Theodore C. *Value Analysis in Design*. New York: Van Nostrand Reinhold, 1990.

15

Design to cost

Design to cost is a method of elevating cost with technical performance so that they are relatively equal. Design to cost is important because it allows planning and control for costs, affordable products, and support for life cycle cost. In this chapter, you'll learn:

- The basics of design to cost.
- Why design to cost benefits you.
- Steps in a design to cost program.

Defining design to cost

The essence of design to cost is making design converge on cost instead of letting the design determine cost. Design to cost is design to objectives, which is the theme of systems engineering. This is different from telling the engineers, "Go make this work, and when you are done, tell me how much it costs."

Design to cost was formalized as costs were increasing, with emphasis on performance and schedule. Engineers were seldom held accountable for costs but always for technical performance. Costs are set by the design, however, and the engineer must be given cost as a design objective.

DTUPC, or design to unit production cost, is a term frequently found with high-volume production. It refers to the recurring cost of the unit product. DTLCC, or design to life cycle cost, includes all elements of the product life cycle. Design to cost can be applied to one delivered product, production runs, and total system and product acquisition.

Benefiting from design to cost

Design to cost allows for planning and control over technical and product costs. It places responsibility for product costs with the design engineers. Design goals for cost must translate into affordable products for the customer.

Implementing design to cost

Design to cost has certain steps in implementation:

- Determine how much the customer is willing to pay for the product and its support. From price, calculate the costs to support that price.
- Establish the top-level cost as an objective and constraint.
- Estimate costs for the product using parametrics, as-likes, or engineering estimates. Use value analysis if the product is like an existing product.
- Sort the cost drivers and dollarize risk in the new design.
- Decide if you will use the estimates or ignore them, hoping for a technical breakthrough. Judgment is critical.
- Establish and allocate target costs.
- Set up the design to cost system. Appoint single-point reporting for design to cost. Define the requirements. Define the communication paths and methods. Determine how trade studies will be performed.
- Use value analysis and producibility methods. Iterate the design and process.
- Review and hold design engineers responsible for meeting cost requirements.

An early opportunity for cost reduction is in functionality versus cost curves. These should be identified and calculated by systems engineering. The customer requirements should be analyzed to find the true performance needs. Small increments of performance frequently mean large increments in cost. Cost drivers can be identified early this way.

Cost drivers can be identified by the use of Pareto analysis. The principle is that 80 percent of your costs are caused by 20 percent of your product. All costs are ranked in descending order of cost. The top items, where efforts have a higher payback, are selected for cost reduction efforts first.

Engineers sometimes have difficulty in estimating costs. This is because of the risk involved. Estimating is usually done at the "should cost" point, which is equal chance for the estimate to be more or less. Large spreads of top and bottom estimates should be reported and incorporated into the risk planning.

Successful design to cost programs have used the following techniques:

- Select a qualified person and let them be the focal point on controlling design to cost.
- Allocate design goals to individuals and hold them responsible.
- Make cost a part of the review process. Have design engineers present cost goals and estimates along with technical parameters.
- Give the design engineers the means to meet cost goals. Provide assistance in producibility and value analysis. Give them access to manufacturing and supplier costs.
- Evaluate alternative approaches and use a quantitative scoring system.

Recordkeeping is essential. Requirements and concepts can change. A baseline is needed to sort cost growth from requirements growth. This is especially true with software.

Summary

- The essence of design to cost is making design converge on cost instead of letting the design determine cost. Design to cost is design to objectives.

Further reading

Michaels, Jack V. and William P. Wood. *Design to Cost*. New York: John Wiley & Sons, 1989.

Part III

Management of systems engineering

Management of Systems Engineering
- Technical Control
- Work Breakdown Structure
- Cost and Schedule
- System Engineering Management Plan
- Proposals

Why is systems engineering important? Systems engineering transforms customer needs and expectations into a product. When products delight the customer, sales provide revenues for the existence of the business. The efficiency of the transformation process is directly tied to product performance, cost, and schedule.

The product made for the customer is the unifying force for the project. People become excited about products that provide a common goal. Profit is not the reason for projects, it is a result of a job done well.

Leadership

In Part III, you'll learn the tools for dealing with complexity. Dealing with complexity is a management function. A project is also about dealing with change, and dealing with change requires more than tools—it requires leadership. Systems engineering is far too interactive and interpersonal to run with tools alone. Successful systems engineering needs both competent management *and* leadership.

But what makes a good leader? In addition to being competent and expert, a good leader has vision and convinces others that his goals are worthwhile. He must have the ability to influence people's acceptance of ideas and shape their concepts for the project. Contributing technically is expected. A good leader must innovate and improve beyond the ordinary.

Ingredients for successful projects

What makes a project a success? There is no proven path to success. Until science discovers the magic formula, perhaps the following list of ideas that seem to work can help:

- Treat others as you wish to be treated.
- Have clear goals.
- Give to receive.
- Respect your team members.
- Acknowledge good work.
- Active participation causes interest.
- Regard people as resources, not expenses.
- Teamwork.
- When goals are clear, understood, and accepted, engineers will self-monitor.
- Communicate a clear value system with priorities and responsibilities.
- It is not enough to set requirements, you must provide the means of meeting the requirements. Training and coaching help people meet goals.
- Reinforce the creative dimension.

System engineering provides the dimensions of logic and information. The dimension of creativity involves all engineers on the project. Engineers are typically self-motivated; they like to solve problems. You must provide a project environment that fosters creativity and encourages it. Today, pleasing the customer is not enough. Tomorrow's successful businesses are already creating what the customer will want three years from now.

16

Technical control

One of the major elements of systems engineering management is technical control. The system engineer takes technical responsibility for the product and project. Systems engineering deals with complexity by establishing a few vital controls. Locations of controls are typically:

- Between major project phases.
- Before irreversible paths, such as before a production go-ahead.
- At vital processes.
- At natural convergence points.

In this chapter, you'll learn about:

- Documenting requirements through specifications.
- Controlling interfaces.
- Conveying work to be done through the statement of work.
- Controlling change through configuration management.
- Controlling risk through technical performance measurement.
- Managing risk.
- Meeting design requirements through technical reviews.
- Determining risk through third-party audits.

Specifications

A specification is a document that establishes technical requirements for performance and design details. It also establishes the means of verification for its requirements. A specification tells what is wanted, not how to do it. Specifications are not statements of work nor do they ask for data. Specifications are important to systems engineering because they establish:

- Requirements.
- Means of verification for requirements.
- An engineering baseline.

Designers without requirements will improvise using past experience and personal feelings. The result is usually less than the customer expected. Poor specifications lead to poor designs. In this section, you'll learn about specification requirements characteristics, writing, and hierarchy.

Specifications are not stand-alone documents. They must be used with the statement of work and engineering plans. For example, the designer needs to know quantities and maintenance concepts that are not part of the specification. A specification does not have the value system or constraints of the manufacturing process either. Systems engineers recognize that specifications do not convey sufficient information for design.

Requirements characteristics

The following are characteristics of well-written requirements:

- Unambiguous
- Complete
- Verifiable
- Consistent
- Modifiable
- Traceable

Unambiguous means that every requirement has only one interpretation. Requirements written in English are susceptible to ambiguity because of word order and emphasis. Formal requirements languages can be used to reduce ambiguity.

Complete means inclusion of all significant requirements, functions, behaviors, performance, constraints, and interfaces. Valid and invalid input values have specified responses.

Requirements are verifiable when a cost-effective means exists for people or machines to check that the product meets the requirement. Approximately, usually, resistant, and fail-soft are not verifiable terms.

Consistent requirements are not in conflict with each other. Different terms, different characteristics, and different sequences are examples of conflict between specification requirements.

Modifiable requirements are easy to change completely and consistently. Modifiable is assisted by an index and cross-referencing, and only states a requirement once.

Traceable means that the origin of the requirement is clear. Upward traceability means that the requirement can be traced through allocation or derivation back to customer needs. Downward traceability finds the impact to lower-level requirements caused by top-level changes.

MIL-STD-490

MIL-STD-490A, *Specification Practices*, describes the preparation, interpretation, and change of specifications. Highlights from this document are the types, language, and sections of specifications.

The three general types of specifications under MIL-STD-490 are types A, B, and C. The B and C types have various subtypes within these categories.

Type A specifications are the system specifications, which:

- State the technical requirements for the system.
- Allocate requirements to functional areas.
- Document design constraints.
- Define interfaces for functional areas.

Type B specifications differs from type A by time. Type B specifications are written after type A. Type B, or development specifications:

- State the requirements for the design or engineering development of a product during the development period.
- Describe the performance characteristics for each configuration item.

Type B specifications include the B1, or prime item development specification, and the B2, or critical item development specification. The key word here is complexity. Prime items are more complex. Critical items are technically complex, reliability critical, or indispensable to the total product.

Type C specifications are product specifications. They are used for any configuration item below the system level. The type C specification covers form, fit, and function.

Specification sections

Specifications have six sections and an optional appendix. The sections are:

1. Scope.
2. Applicable documents.
3. Requirements.
4. Quality assurance.
5. Preparation for delivery.
6. Notes.
10. Appendix.

The scope section is normally a brief overview that describes the product and what it is supposed to do. The scope should contain enough information for the reader to understand the intent of the product.

The applicable documents section is a listing of all referenced documents in the specification. The revision or date of all documents must be explicitly stated.

The requirements section provides all technical requirements. The quality assurance section provides for the verification of section three requirements. Figure 16-1 illustrates the correlation of section three requirements with section four verification. Section three and four paragraph numbering follows a standard scheme for a specification type in MIL-STD-490.

Instructions for the preparation and physical delivery of the product are in section five. Nontechnical clarifications or other general information is included in section six.

Writing a specification

When writing a specification, the language style should:

- Be free of vague and ambiguous terms.
- Use the simplest words and phrases to convey the intended meaning.
- Include complete essential information.
- Be consistent in terms and nomenclature.
- Use short, definitive statements.
- Use the emphatic form of verbs ("The device shall weigh . . .").

Rules for using "shall," "will," "should," and "may" are:

- Use "shall" to express mandatory provisions.
- Use "should" and "may" to express nonmandatory provisions.
- Use "will" to express declaration of purpose on the writer's part or simple future tense of verbs.

Use these words to cite referenced documents:

- "conforming to".
- "as specified in".
- "in accordance with".

Corrections or updates to specifications must be handled by a formal change notice. Number and date all pages of a specification to facilitate configuration control. If you replace one page with more than one, additional pages are numbered the same number with the addition of a suffix letter starting with "a." For example, 4, 4a, 4b, etc.

As you outline and write the specification, ask yourself:

- Is the requirement necessary?
- Is the requirement overspecified?
- How can the requirement be verified?
- Is there another way to interpret the wording?

Specification hierarchy

Specifications follow a hierarchy that reflects the physical system. A specification tree acts as a road map to project specifications. At the top system level, a functional specifi-

Technical control

DESIGN SHEET	NOMENCLATURE COMMUNICATIONS CONTROL PANEL (CCP)	CEI NO. OR CRITICAL COMPONENT CODE IDENTIFICATION 345678A Detail Spec No. _____

Requirements for Design and Test
Verification Cross-Reference Index

Method legend: N.A. - Not applicable   2 - Review of analytical data
               1 - Inspection   3 - Demonstration   4 - Test

Cat. I Legend: A-Engineering test and evaluation   B-Preliminary qualification
               C-Formal qualification   D-Reliability test and analysis
               E-Engineering critical component qualification

Section 3.0 Requirement Reference	Verification Method(s)					Test Category						Section 4.0 Verification Requirement
	NA	1	2	3	4	A	B	C	D	E	II	
3.0	x											
3.1	x											
3.1.1				x				x				4.1.3.3a
3.1.1.1				x				x				4.1.3.3b
3.1.1.2				x				x				4.1.3.3c
3.1.2	x											
3.1.2.1			x						x			4.1.3.2, 4/1.4
3.1.2.2	x											
3.1.2.2.1				x				x				4.1.3.3d
3.1.2.2.2				x				x				4.1.3.3e
3.1.2.2.3				x				x				4.1.3.3f
3.1.2.3	x											
3.1.2.4	x											
3.1.2.4.1					x			x				4.1.3.4
3.1.2.4.2					x			x				4.1.3.4
3.1.2.5				x				x				4.1.3.3g
3.1.2.6		x						x				4.1.3.1
3.1.2.7	x											
3.1.2.7.1		x						x				4.1.3.1
3.1.2.7.2	x											
3.1.2.8				x				x				4.1.3.3h
3.2	x											
3.2.1		x						x				4.1.3.1
3.2.1.1		x						x				4.1.3.1
3.2.1.2		x						x				4.1.3.1
3.2.2	x											
3.2.2.1	x											
3.2.2.2	x											
3.2.2.3	x											

Revision __B__   Approval _____   Date __4 Jun 1965__   Page No. __12__ of __13__

Fig. 16-1. Verification cross-reference index example. (Source: AFSCM 375-5)

188 Management of systems engineering

cation is prepared first. Later in time, but at a system level, an allocated requirements specification is prepared. This is a "design to" specification, which is the basis for later "build to" specifications and drawings. From the system level, requirements flow down to the subsystem level. Specifications flow down to the component level, which is a functional unit such as an electronic board or software module.

The role of the system engineer changes as the flow down process proceeds. The system engineer probably writes the top-level specifications. Because of the huge work load, lower-level specification writing assignments are given to the lead engineers and designers. Systems engineering approves those specifications, however.

At the lower level of a product, design sheets are sometimes used instead of specifications. Design sheets can be used to contain additional internal documentation not suited for specifications. Design sheets contain both a section three "Requirements" and a section four "Quality Assurance Provisions." When appropriate, the function source reference number is entered to the left of the applicable requirement. Design sheets describe the intended design approach in addition to the requirements. The design sheet is basically only the requirements and test sections for lower-level components.

Commercial specifications

Commercial product design specifications typically include such areas as:

- Standards and specifications.
- Interfaces.
- Environment—use, manufacturing, storage, transportability. etc.
- Service life.
- Maintenance.
- Size.
- Weight.
- Power.
- Appearance and finish.
- Materials.
- Ergonomics.
- Reliability.
- Safety.

Writing for procurement

Specifications written for procurement must be used with a purchase order. Your objectives for the specification and purchase order are to:

- Get the work done on time, to specification requirements, and at the lowest cost.
- Provide a baseline so both you and the seller know when the work is done.

Defective specifications prevent performance by the seller. A specification might be defective through:

- Errors, omissions, or conflicts.
- Buyer-specified solutions that don't work.
- Brand name requirements when the brand name is not obtainable.
- Failure to disclose superior knowledge.

In commercial procurement, the law recognizes impossibility and impracticality. Contracts that become substantially more difficult to perform than envisioned by the parties at the time of execution can lose enforcement by the law.

Interfaces

An interface is the functional and physical characteristics required to exist at a common boundary between two or more equipment items or computer programs. Interfaces result from partitions that create boundaries. Partitions can be external, internal, or work-related partitions.

Partitioning should be logical. The logic might follow functional allocation, or it could be sequential, where the logic is data flow.

Partitioning should minimize the interaction across boundaries. For example, an excessive number of wires between boxes suggests poor minimization of coupling and poor cohesion within the boxes. Other considerations in partitioning are:

- Testability.
- Maintainability.
- Reliability.
- Safety.
- Ease of manufacturing.
- Incompatible materials.
- Robust inputs and target outputs.
- Normal and unnormal conditions.
- Packaging, transporting, and storing.

Testability and maintainability considerations are obvious in interfaces. Reliability can best be illustrated through aircraft. On some aircraft, 60 percent of all failures are caused by connectors or wiring. By reducing the wiring across interfaces, reliability could improve significantly.

High voltage and high currents at interfaces affect safety. Where possible, avoid these conditions. If not possible, human exposure to contact must be prevented. Accidental shorting by accident, testing, or environment must be prevented by design.

Interfaces should be robust to changes and noise in the inputs. Normal and out of normal conditions must be considered at interfaces. Outputs should be held close to target conditions. Interactions at all interfaces should be explicit and obvious. Interface control documents control the requirements for interfaces. Write the documents for four reasons. To:

1. Establish compatibility between items.
2. Control interfaces early to minimize changes.
3. Communicate with all project team members.
4. Prevent "I'm waiting on so-and-so to begin my design."

Statement of work

The statement of work describes tasks, products, and services. Accurately conveying this information scopes the work to be done. In this section, you'll find:

- A description of the statement of work.
- How the statement of work and specification go together.
- How to prepare the statement of work.

Describing the statement of work

The statement of work (SOW) establishes tasks and identifies the work effort to be performed in terms of minimal needs. The statement of work defines the work tasks that must not be contained in a specification and that are not in the data requirements list. Elements usually found in a statement of work are:

- The scope of work-including objectives.
- Tasks.
- Deliverable end items.
- Support equipment.
- Support services.

Relation to specification

The statement of work is not a stand-alone document. The system specification identifies performance and supportability requirements for the system. The statement of work does not address performance characteristics. It is supposed to task activities.

Preparing the statement of work

Before writing the statement of work, understand completely what you want or require. Review the systems engineering documentation to date. Identify potential cost drivers and include only the minimal necessary needs. Prepare a preliminary work breakdown structure. Identify all people who will contribute in preparing the statement of work. Above all, specify what you require, not how it should be accomplished.

After talking with the specialists and concerned people, divide the work into logical groupings. Develop an outline of how the tasks will be covered. Single out work that will require special care because of technology or risk. Form a team and give assignments.

A title page and table of contents begin the statement of work. Following those is a standard format of three sections:

1. Scope.
2. Applicable documents.
3. Requirements.

The Scope section includes a brief statement of what the statement of work covers. It might include an introduction and background.

Applicable documents invoked by specific reference in the statement of work test are listed in section two. This enables the reader to obtain the documents and understand fully what you need. Section two should not be prepared until after section three is complete.

Section three describes the requirements. Group the work or task efforts logically. Chronological order is desirable. Requirements can be mandatory, desirable, optional, or have alternatives. These are designated with the proper language of shall (mandatory), should, or may.

Table 16-1. Timing of Reviews.

Review	Looks at	When
Systems Requirements Review (SRR)	Technical requirements definitions Support systems analysis Engineering management	Significant portion of the system's functional requirements are established
System Design Review (SDR)	Review completed system specification Allocated technical requirements Risk Engineering planning for next phase System engineering process	System characteristics are defined Allocated configuration identification is established
Preliminary Design Review (PDR)	Each configuration item: Evaluate progress, technical adequacy, risk of design approach Determine compatibility with requirements Complete interfaces	After development specifications are written but before start of detailed design
Critical Design Review (CDR)	Meets requirements Detail design compatibility with support and personnel Producibility Risk	Detail design is essentially complete

Use simple words and concise sentences. Avoid abbreviations and jargon as much as possible. Use active verbs rather than passive. Recognize that the reader might be an engineer, buyer, accountant, lawyer, cost estimator, or other person of different background. Use the work words of MIL-HDBK-245B as guidance.

Include procedures. When decisions have not been made, include the procedure for making them. For example, "as approved by the buyer," or "submit to be approved or disapproved by the buyer within 30 days."

Ask yourself these questions as you write the document:

- Is the statement of work specific enough to determine cost, expertise, labor, and other resources required?
- Are the tasks stated so a contracts person can determine compliance?
- Are sentences stated in terms of "you shall do this work" and not "this work is required?"
- Are work tasks presented in chronological order?
- Can the reader distinguish directions from general information?
- Are proper quantities shown?

Configuration management

Configuration control is important for:

- Keeping form, fit, and functions.
- Controlling interfaces.
- Controlling safety characteristics.
- Maintaining records of changes.

Configuration management includes:

- Configuration identification—selection of the documents that identify and define the configuration baseline characteristics of an item.
- Configuration control—controlling changes to the configuration and its identification documents.
- Configuration status accounting—recording and reporting the implementation of changes to the configuration and its identification documents.
- Configuration audit—checking an item for compliance with the configuration identification.

Configuration change control ensures that:

- Each proposed change, approval of change, and documentation prior to implementation is properly evaluated.
- Authorized changes are accomplished.
- Functional and physical interfaces are controlled.

Interface control is a part of configuration control. Interface control identifies:

- The parties affected.
- The methodology for establishing and controlling interfaces.
- The establishment of interface control working groups (ICWG).
- ICWG meetings.
- Coordination and processing of changes.
- Identification of the basic types of documents to be used.

Major control points are called baselines. Baselines establish configuration definitions from which configuration changes can be tracked and documented. The three main baselines for products are functional, allocated, and product.

The functional baseline:

- Defines user and technical requirements.
- Allocates requirements to functional areas.
- Documents design constraints.
- Defines interfaces.

The allocated baseline:

- Details design requirements.
- States performance characteristics.

The product baseline defines:

- Form—proper physical characteristics.
- Fit—correct interfaces.
- Function—correct operating characteristics.
- Performance and test requirements for acceptance.

Configuration accounting at the product baseline could include facts such as:

- The physical item (manufacturing sequence number 090) conforms to revision B of the engineering drawing.
- The integrated circuits were purchased from Acme Supply on purchase order 910788.
- Assembly was performed by operator 107 on April 15, 1991 and inspected by 899 on the same day.

Permission for nonconformance with baseline configurations is granted through deviations or waivers. A deviation is permission before building a configuration item to not comply. Deviations can overcome temporary conditions or can be used for proof of concept test items. Deviations are temporary, because the configuration item documentation is not changed. Waivers are permission for nonconformance after building the item under configuration. Waivers are not planned and are not desirable.

194   Management of systems engineering

The change control board (CCB) has charge of all change control activities. The CCB evaluates changes for potential effects on hardware, software, interfaces, testing, training, manuals, spares, cost, schedule, reliability, and other areas. The change control board should be chaired by the system engineer or other responsible person. Members typically include the:

- System engineer
- Project manager
- Design engineering
- Test engineering
- Reliability
- Maintainability
- Manufacturing
- Quality assurance
- Configuration management
- Logistics

Changes are sometimes classified by their degree and priority. For example, a Class A change can be anything that affects form, fit, function, or a specification requirement. A Class B change would be minor changes, such as typos. Some organizations give special priority to changes that affect safety. Document these decisions in the configuration management plan.

The configuration management plan is a part of systems engineering. Systems engineering identifies the configuration items of the system. The plan documents decisions about controlling conformance to requirements and revisions to documentation and the system.

A competent configuration manager performs a specialist role for the project. The configuration specialist provides expert advice and assistance in preparing the configuration management plan, for example. A configuration team should record and manage the configuration paperwork for your project.

Technical performance measurement

Technical Performance Measurement (TPM) is the product design assessment that estimates or measures the values of essential performance parameters. TPM:

- Forecasts values to be achieved.
- Measures differences between allocated and achieved values.
- Determines impact on product effectiveness.

Technical performance measurement is important because it is the technical control corresponding to cost and schedule controls. In this section, you'll learn how TPM fits in the System Engineering Management Plan and what elements are in TPM.

Technical performance measurements are used to evaluate the impact to cost, schedule, and technical effort. It is useful for revealing unwanted deviations to product performance. TPM is another means to increase your confidence in the design effort. Technical performance measurement is a part of risk management.

Writing TPM into the SEMP

The technical measurement performance section of the System Engineering Management Plan (SEMP) includes:

- The objectives of each assessment.
- Scheduled times.
- Selection of performance parameters.
- Forecasted values.
- The methods and conditions of assessment.
- Identification of data required.
- The acquisition of data.
- The quantitative and time-phased expected results.

The steps of a performance measurement system are:

1. Establish a measurement system with milestones and verification plans.
2. Measure results.
3. Assess deviations.
4. Report results.
5. Forecast deviations and potential risks.
6. Take corrective action.

The measurement system must be both useful and usable. Don't measure what you won't analyze. It can be helpful to diagram a performance tree that breaks down and finds critical parameters. Select parameters by:

- The most significant determinants of technical product performance.
- Their ability to directly measure value from analysis or demonstration.
- Their ability to predict and verify time-phased values.

One word of warning. TPM is susceptible to the 90 percent complete syndrome. Graphs especially will move to the target value early. Graphs should not be shown to the data providers. Data must be verified. The setup of the measurement system is extremely important to its success.

Managing risk

Risk has two components: likelihood and severity. Likelihood is the probability that the loss will occur. Severity is the full consequence of the loss. For example, earthquakes can be severe, but the likelihood of having one is very low for most places. Risk management has assessment and mitigation. The steps for managing risk are to:

- Identify potential loss.
- Examine the likelihood and severity of the loss.
- Identify likely causes of the loss, for high risks.
- Identify specific preventive actions to remove or reduce the likelihood of the causes.
- Plan contingent actions to minimize the loss.
- Identify decision rules to activate contingent actions.

To get resources to solve the risk problems, you must speak the universal language of managers: you must dollarize the risk. Calculate the severity of the loss should it occur, then multiply by the probability of that event happening. This gives you the dollarized risk. Add all of your risks into a total amount. This is the risk reserve the project needs to carry. The risk reserve must be calculated periodically because the risk changes with problems solved and new problems appearing. Look for risks in such areas as:

- Critical and near-critical path.
- New technology.
- New process.
- Interfaces.
- Poorly defined requirements.
- Work packages using scarce resources.
- Work packages with many feed-in or feed-out precedences.
- Work packages with people poorly fitted to the job.

A well thought-out work breakdown structure and a systematic review of each element is one way to find risk. Technical performance measures and measures of effectiveness are others. A risk management plan might include:

- Risk identification methods.
- Risk quantification.
- Risk handling.
- Risk tracking.
- Risk budgeting.
- Contingency planning.
- A technical risk summary.

Risk management must be on-going throughout the life of the project. Each identified risk belongs on its own risk sheet, kept in a risk book. Prepare a watch list from the book of risks and contingency decision points. Work the highest risks first until each risk is resolved.

Technical reviews

Reviews are structured presentations of technical information that you prepare. In an audit, the auditor has a checklist of what the auditor wants to see. Technical reviews and walk-throughs are important because they catch errors earlier when it is less expensive to change them, and speed design through communication. In this section, you'll find:

- Review objectives.
- Characteristics of good reviews.
- Planning reviews and walk-throughs.
- Conducting reviews.

Review objectives

The objective of a review is to ensure that the product meets the requirements. Secondary goals in support of this are to:

- Evaluate the design for compliance.
- Ensure that the product is safe and reliable.
- Verify interfaces compatibility.
- Prevent mistakes and omissions.
- Challenge the design for optimization.
- Assess risks.

Setting goals for the review

Goals should be set in review preparation. Goals might include:

- A statement of review needs.
- The expected results and actions from the review.
- The length of the review, level of detail, and how much documentation to provide.
- The desired attendance.
- Criteria to measure acceptability.
- An agreed upon agenda.

Each presenter must be given standards to prepare against. You must communicate the order of presentation, topics, time allotted, and number of visuals. An outline will save both the presenters and the project time and money. Plan your reviews.

Generally accepted terms for reviews include the system requirements review, the systems design review, the preliminary design review, and the critical design review. Systems engineering can influence the design most at the earlier reviews. The characteristics and timing of these reviews are in TABLE 16-1. Checklists for reviews are in MIL-STD-1521B, *Technical Reviews and Audits for Systems, Equipments, and Computer Software*, located in the appendix.

198 Management of systems engineering

| PROJECT | RISK TRACKING FORM | DATE   REV |
| | ID#   STATUS | SHEET   OF |

| WBS NUMBER | ELEMENT TITLE |

Potential Problem

Description:

Likelihood:
Severity:

Likely causes:
1.

Preventive actions for likelihood reduction:
1a.

Contingent actions for alleviation
A.

Decision points for contingency action
A.

Fig. 16-2. Risk sheet example.

Walk-throughs

Walk-throughs are informal, interactive reviews. Hold walk-throughs at logical progress points in the development phase. Walk-throughs can be more effective than reviews because they happen during the development phases and not after completion. They also are more honest, as reviews tend to be "dog and pony show" when management or customers attend.

Walk-throughs should be done when entry criteria are met. Designers should have entry criteria communicated to them so that they can have their own walk-throughs. Walk-through announcements have the reviewers' names, the subject, time, and place. Attendees should be held to a minimum. Walk-throughs last for about an hour, never more than two hours. Systems engineering receives a copy of the action list prepared during the meeting.

A trained moderator and recorder can facilitate walk-throughs. The recorder keeps notes in the form of an action list. The moderator keeps the meeting running and on technical issues, not personalities. Issues are raised by the reviewers, but problems are solved later by the person being reviewed; not during the meeting. Resolution of action items are documented by the reviewee and approved by a responsible design leader.

Resolving problems

Resolving problems usually means further design iteration. Alternatives include:

- Reverifying design adequacy through analysis or empirical proof.
- Redesigning to correct problems.
- Reallocating design requirements to correct allocation errors.
- Redefining design requirements to move to an adjacent solution space.
- Reevaluating customer requirements to fall within constraints.

Effective design reviews

Design reviews become ineffective when they become:

- Milestone oriented instead of design-criteria oriented.
- Success oriented, not technically oriented.
- Superficial—too little time, poorly organized, poorly prepared.
- Careless—poor recording of results and action items.

To have effective design reviews:

- Be prepared.
- Raise issues, don't solve them in the meeting.
- Stick to the technical.
- Keep out onlookers and straphangers.
- Record issues and problems.

Conducting reviews

There are more logistics than you might first think. Planning includes:

- Room readiness (always check the room).
- Breaks.
- Meals.
- Quorum time for the reviewers.
- Recorder or note-taker for errors and issues.

Managing the speakers requires you to:

- Set objectives.
- Allocate time.
- Plan the information you want delivered.
- Sequence the speakers.
- Introduce the speakers.

Delivering the information can be organized:

- From general to specific.
- Input to output.
- From situation to problem to solution.

Visual aids must:

- Be readable at the back of the room.
- Have only one theme per visual.
- Have no more than six "bullets."
- Have no more than six words per bullet.
- Never be read out loud by the presenter.

Reading your visuals is boring to the audience. In a way, it also implies that they are not smart enough to read. Develop a script that supports the points you are making with your visual.

Reviewing your suppliers

Reviewing your suppliers takes judgment. They are likely to have more depth in their particular area than you. You can be swamped by information overload in a short review. You can only concentrate on broad issues such as:

- Are the requirements correct, and is there a flow-down?
- What is the value system?
- Are alternatives presented with trade studies?
- Does the risk management appear reasonable?

- Are specialty and design to cost goals allocated to the designers? Are they tracking progress against goals?
- Is enough time being allotted to the review?
- Was a functional analysis performed? Did you see a functional flow block diagram or equivalent?
- Are all the interfaces identified and defined?
- Is the review happening because design criteria were passed or because payments are tied to the review as a milestone?
- Is life cycle cost given importance?
- Are normal and out of normal conditions considered?
- Are environments considered by the designers?
- Does it appear that they will deliver on time?
- Are you receiving straight answers to your questions, or do they behave defensively? Is the team happy or burnt-out?

Auditing systems engineering

To independently determine risk and potential trouble for the project, a systems engineer can ask for an audit. Management calls for most audits today after a project is already in trouble. Audits performed in the front third of the estimated project duration are the most useful. Problems found earlier can be corrected with the least cost.

In this section, you'll learn:

- What the responsibilities of the auditor are.
- How to perform an audit.
- Measuring.
- Reporting principles.
- How to write the audit report.

Responsibilities of the auditor

The auditor formally compares the performance of the system engineering team to the requirements. An audit is conditional on a clear assignment of responsibilities and firm objectives for the project. The auditor investigates the:

- Documentation and customer needs.
- Tangible results of progress.
- The system engineering process.
- The system engineering management.
- Planning and controls.
- Conformance to internal and external standards and procedures.

From the investigation, the auditor reports to meet these objectives:

- Current status compared with obligations and expectations.
- Forecast status at future times compared with obligations and expectations.

- Critical risk issues.
- Critical system engineering management issues.

The auditor must determine the audit requirements and develop realistic expectations by the requester of the audit. Although the auditor must be competent in systems engineering, he or she cannot be expected to be expert in all areas. When an opinion is needed, and the auditor is not certain, the auditor must state a lack of qualification in that area and decline comment.

The auditor keeps the report confidential. The system engineer and the requester of the audit receive copies and control distribution. The auditor does not discuss the findings nor make them public. Opinions expressed in private to the auditor are investigated for confirming facts. The auditor does not expose confidential sources.

The auditor maintains independence in all matters of the audit, including confidentiality of the report. The opinion must be independent, without bias or coercion.

Confirmation is the basis for the opinion. An audit is an act of measurement; it is not a review or walk-through. The audit is an objective statement of fact that reduces uncertainty. Using rumors, information, and remarks without confirmation destroys audits. It is unethical and possibly harmful to the people under audit. Confirmation means:

- Information coming preferably from two independent sources.
- A logical way of validating or deriving the facts from data.
- Supporting opinions in the report with confirmed facts.

Performing an audit

The auditor must have access to all records that allow determining progress and future progress. Records might include, but are not limited to:

- Background information on customer and customer needs.
- Telephone and meeting reports with the customer.
- Internal memos.
- Contract, statement of work, specifications, and data item lists.
- List of all deliverables with dates.
- System Engineering Management Plan.
- Project schedules.
- Standards and policies.
- Status reports.
- Personnel qualifications and experience.
- Labor time summaries.
- Interface definitions.
- Trade studies reports.
- Work breakdown structures and work packages.
- Critical path.
- Staffing plan.
- Resource loading.

- Integration plan.
- Risk management documents.
- Cost estimation and actuals.
- Mission profile.
- Functional analysis.
- Verification plan.
- Requirements traceability.
- Design to cost plan.
- Purchase order status.
- Project design guides.

Working papers organize this information. Filing and cross-referencing keeps information usable and retrievable. Working papers set up an organized approach towards the audit objectives. They provide traceability and a way of recording issues for later investigation. Individual papers must be numbered to facilitate this. On a large audit, working papers allow more than one person to use the information.

Working papers include the open questions, suspected risk list, critical issues, and schedule of the audit. The auditor should keep an organized workbook with forms and checklists for the audit.

Performing the audit requires tact and discretion. Politics will enter any audit situation. The auditor can even be viewed as a management threat. The auditor must maintain independence and detachment from the political arena.

If the auditor interviews the customer, extreme sensitivity must be exercised. The auditor must make no comments about the findings, even if they are requested. It is easier when the customer is not aware that an audit is taking place. The auditor must let the audit requester and the project manager retain control during periods of direct customer contact.

Any interview should respect the time of the person under interview. The auditor asks questions in a respectful way. The auditor never gives the impression that he or she thinks they can know more about the project than those working on it. The auditor uses good interviewing skills to draw out relevant information.

Measuring

Valuating verifiable deliverables is the basic approach to measuring. The valuation basis is that work is of value only if it reduces the amount of work to be done. Again, money spent is not a measure of progress.

Deliverables must be something that exists and can be analyzed and evaluated. Deliverables are both internal and external items. Work must be in a tangible form to be measured by an auditor.

Milestones are progress measurement points in time. Milestones can be considered tangible deliverables with associated completion dates. A value can be assigned to completed milestones.

Assigning value is different from looking at cost. If the work package has exceeded its budget and work remains, cost measurement has identified the risk after the fact. It is difficult to assign partial value using a consistent and repeatable method. If a task is in a middle stage, is progress happening? Will the task complete faster or slower than forecasted? The auditor determines the value of work done that reduces the amount of work to be done.

Reporting principles

Accountants have general principles for opinions and reporting. These include relevance, verifiability, objectivity, consistency, comparability, materiality, and disclosure. Each of these words conveys considerable meaning to an accountant.

Borrowing from accounting theory, systems engineering auditing can follow certain principles also. Applicable principles include relevance and materiality. The audit discloses important and relevant facts but also considers their weight and significance. The decision about what is important is somewhat a matter of the auditor's judgment and experience.

All opinions and information in the audit report must be verifiable. The report reader assumes that all facts have been confirmed. The objective of the audit is to reduce uncertainty.

A major reason to request an audit is to have an objective opinion. The report must be based on actual evidence that a third party could confirm. Actual evidence is not based on statements from the project team but on facts. The auditor must remain impartial.

Audits are measurements against checklists and predefined standards of performance. Without these, the audit is a personal expression of the auditor. The auditor must follow the principle of consistency in reporting each project.

The principle of disclosure means that all relevant facts are disclosed. All data needed to understand and interpret opinions must be included. Hiding embarrassing facts will only make matter to worse as the project heads into deeper trouble.

Writing the audit report

The auditor writes the report using the working papers. The organization of the working papers expedites the writing process. The auditor gathers needed information by performance standards and checklists. The major headings of an audit report are:

- Introduction.
- Current status.
- Cost performance (reflecting consumption of resources).
- Schedule performance.
- Progress performance (valuation of work done).
- Quality and compliance performance.
- Forecast of future performance.
- Critical issues.
- Risk assessment.

The auditor develops these sections in detail according to the level of audit requested. For example, in the quality section, significant issues would be anything that would cause the product to be rejected.

In most technical reports, there is a section for recommendations. There are two views on including recommendations in an audit. One is that, if an auditor is to report impartially, then recommendations are improper. The second pragmatic view is that a systems engineer competent to perform audits should recommend improvements. Because most companies perceive a lack of qualified systems engineers, the auditor will probably have to make recommendations about the problems uncovered.

Summary

- A specification is a document that establishes technical requirements for performance and design details. It also establishes the means of verification for its requirements.
- Specifications are not stand-alone documents. They must be used with the statement of work and engineering plans. Well-written requirements are:
  - ~ Unambiguous
  - ~ Complete
  - ~ Verifiable
  - ~ Consistent
  - ~ Modifiable
  - ~ Traceable
- MIL-STD-490A, *Specification Practices*, describes the preparation, interpretation, and revision of specifications.
- As you outline and write the specification, ask yourself:
  - ~ Is the requirement necessary?
  - ~ Is the requirement overspecified?
  - ~ How can the requirement be verified?
  - ~ Is there another way to interpret the wording?
- Partitioning should minimize the interaction across boundaries.
- The statement of work describes tasks, products, and services. Elements usually found in a statement of work are:
- The scope of work-including objectives.
  - ~ Tasks.
  - ~ Deliverable end items.
  - ~ Support equipment.
  - ~ Support services.
- Configuration control is important for:
  - ~ Keeping form, fit, and functions.
  - ~ Controlling interfaces.
  - ~ Controlling safety characteristics.
  - ~ Maintaining records of changes.
- Major control points are called baselines. The three main baselines for products are functional, allocated, and product.
- Technical Performance Measurement (TPM):
  - ~ Forecasts values to be achieved.
  - ~ Measures differences between allocated and achieved values.
  - ~ Determines impact on product effectiveness.

- Risk has two components: likelihood and severity.
- Risk management has assessment and mitigation.
- The objective of a review is to ensure that the product meets the requirements. Secondary goals in support of this are to:
  ~ Evaluate the design for compliance.
  ~ Ensure that the product is safe and reliable.
  ~ Verify that interfaces are compatible.
  ~ Prevent mistakes and omissions.
  ~ Challenge the design for optimization.
  ~ Assess risks.
- Walk-throughs are informal, interactive reviews. Walk-throughs can be more effective than reviews because they happen during the development phases and not after completion.
- Resolving problems usually means further design iteration. Alternatives include:
  ~ Reverify design adequacy through analysis or empirical proof.
  ~ Redesign to correct problems.
  ~ Reallocate design requirements to correct allocation errors.
  ~ Redefine design requirements to move to an adjacent solution space.
  ~ Reevaluate customer requirements to fall within constraints.
- The auditor formally compares the performance of the system engineering team to the requirements. The auditor investigates the:
  ~ Documentation and customer needs.
  ~ Tangible results of progress.
  ~ System engineering process.
  ~ System engineering management.
  ~ Planning and controls.
  ~ Conformance to internal and external standards and procedures.

Further reading

Adamy, David. *Preparing and Delivering Effective Technical Presentations*. Norwood, MA: ARTECH HOUSE, 1987.

Boehm, Barry W., ed. *Tutorial: Software Risk Management*. Washington, D.C.: IEEE Computer Society Press, 1989.

Defense Systems Management College. *Risk Management: Concepts and Guidance*. Washington, D.C.: U.S. Government Printing Office, 1989.

Eggerman, W. V. *Configuration Management Handbook*. Blue Ridge Summit, PA: TAB Books Inc., 1990.

Freedman, Daniel P. and Gerald M. Weinburg. *Handbook of Walkthroughs, Inspections, and Technical Reviews*. 3rd Ed. Boston, MA: Little, Brown and Company, 1982.

U.S. Department of Defense. *Preparation of Statement of Work (SOW)*. Military Handbook 245B. Washington, D.C.: U.S. Department of Defense.

―――. *Specification Practices*. Military Standard 490A. Washington, D.C.: U.S. Department of Defense.

―――. *Technical Reviews and Audits for Systems, Equipments, and Computer Software*. Military Standard 1521B, Notice 1. Washington, D.C.: U.S. Department of Defense.

17

Work breakdown structure

A Work Breakdown Structure (WBS) is a product-oriented family tree of the work to be done on a project. The Work Breakdown Structure establishes a logical indentured framework for correlating schedule, cost, technical performance, and technical interfaces. It ensures that planned efforts contribute to project objectives. Preparing a Work Breakdown Structure will help you to:

- Provide a breakout of small tasks that are easy to:
  - understand
  - schedule
  - estimate resources
  - estimate staff
  - assign responsibility
- Ensure that all required tasks are included without duplicating work.
- Organize the work.
- Control the progress of the work by providing a baseline.

In this chapter, you'll learn how to benefit, develop, and change a WBS.

Defining terms

A Work Breakdown Structure is a product-oriented family tree division of hardware, software, services, and project-unique tasks that organize, define, and graphically display the product to be produced, as well as the work to be accomplished to achieve the specified product.

Figure 17-1 shows a tree format for a Work Breakdown Structure. Figure 17-2 illustrates the same WBS in outline format. Figure 17-3 is a WBS in work for a garage door opener.

A Work Breakdown Structure Element is a discrete portion of a Work Breakdown Structure. An element can be an identifiable product, a set of data, or a service.

A Work Breakdown Structure Dictionary is a document that briefly describes the

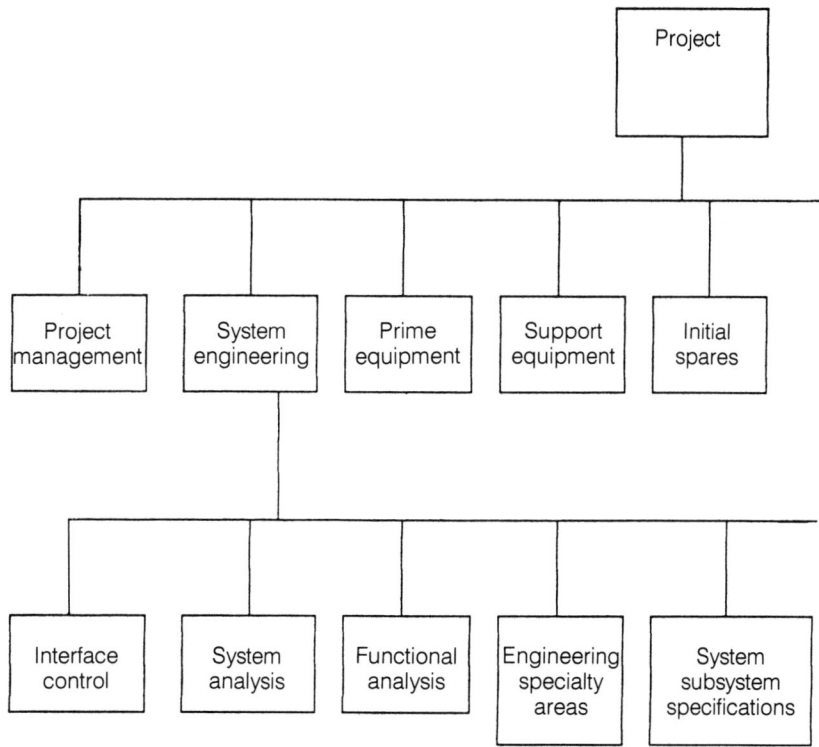

Fig. 17-1. Tree format for a WBS.

Project
    Project management
    System engineering
        Interface control
        System analysis
        Functional analysis
        Engineering specialty areas
        System and subsystem specifications
    Prime equipment - Design and development
    Support equipment - Design and development
    Initial spares

Fig. 17-2. Outline format for a WBS.

tasks of Work Breakdown Structure elements in product-oriented terms.

A Work Package is a detailed, short-span task or material item required to support success of the project objectives.

Work Package Budgets are resources formally assigned to accomplish a work package. They are expressed in dollars, hours, standards, or other definitive units.

Preliminary
Garage door opener project (development)
    Garage door opener
        Opener
            Motor
            Drive mechanism
            Electronic controller board
            Mechanical assembly
            Wiring harness
        Remote control receiver
            Custom integrated circuit
            Circuit board
        Remote control transmitter
            Custom integrated circuit
            Circuit board
    Training
        In-house service technicians
            Training manual
            Place to do it
            Instructors
        Distributors service technicians
            Training manual
            Place to do it
            Instructors
        Technical sales reps
            Training manual
            Place to do it
            Instructors
Test equipment
    Motor test set
    Controller board test set
    Remote control receiver test set
    Remote control transmitter test set
    Wiring harness test set
    Purchased radio test set
    Purchased multi-meters
Systems test and evaluation
    Independent lab testing
    FCC testing
    Safety testing
    UL testing
    Life testing for warranty data
    Climatic tests
System/program management
    System engineering
    Product specification
    Design to cost
    Human factors
    Safety
    Patent searches
    Decide on worm or chain drive
    Project management
Data
    Technical publications
    Owners manual
    Repair manual
    Dealer's installation manual
    Engineering data
    Management data

Fig. 17-3. WBS example for a garage door opener.

Benefiting from a WBS

The Work Breakdown Structure is a tool, an intermediate step to an end. By itself, it has no value other than helping you understand the work to be done. Its value is in organizing information for defining a:

- Critical path
- Schedule
- Risk
- Staff
- Responsibility
- Resources
- Budgets

Work packages compose the WBS. Work packages contain tasks with the resources necessary for the tasks. The structure of the WBS serves as the framework for collecting and computing the costs and resources consumed. It breaks down the complexity, allowing summation of each level of subdivided elements. Work Breakdown Structure elements are selected to structure budgets and control resource use.

After work packages are identified, the order of precedence for their execution must be assigned. The Work Breakdown Structure does not show precedence. After identifying all work packages and their precedence, a critical path can be calculated. The critical path tells you both schedule and risk. Consider tasks on or near the critical path schedule risks.

The summation of all work packages is the summation of people, resources, and money needed. The resources time-phased chart of use is the resource loading. Project scheduling requires a second pass to level resources. Allocating one person for three months of work in one month is not practical, for example. The scheduling process is recursive to balance the duration of the project versus available resources. The raw data for these calculations comes from the structured work packages.

Developing a WBS

A common error in developing a Work Breakdown Structure is to subdivide the work according to people—functional organization. Figure 17-4 shows the matrix formed by the Work Breakdown Structure and the organizational breakdown structure (OBS). A summary of the person or functional group responsible for each work package is commonly called a responsibility assignment matrix (RAM). Patterning the Work Breakdown Structure after the people organization will hide the true costs of work performed. For example, if maintainability efforts are all in one work package, there is no way to tell how much of that effort went into each component or subsystem. The costs for hardware or software development would be understated. Properly coding the cost charge numbers will allow tracking both at the work element and at the functional organization.

Engineering and manufacturing charge numbers are usually different. A work package can have different charge numbers associated with it for engineering labor, manufacturing labor, and materials. If charge numbers are split among or between work packages, then you have a problem. You cannot determine where work is really occurring. It is all right for a work package to have several charge numbers or work orders, but it is not all right for a charge number or work order to cover several work packages.

Organizing the work

A Work Breakdown Structure book saves time during WBS development and use. The book should:

- Explain terminology.
- Show the structure as a tree or indented outline.
- Explain coding.
- Contain the dictionaries in numerical order.

The dictionary is a document that briefly describes the tasks of Work Breakdown Structure elements. Figure 17-5 shows an example of a dictionary sheet. Dictionary sheets are usually prepared before work packages, but the process may be recursive.

212 Management of systems engineering

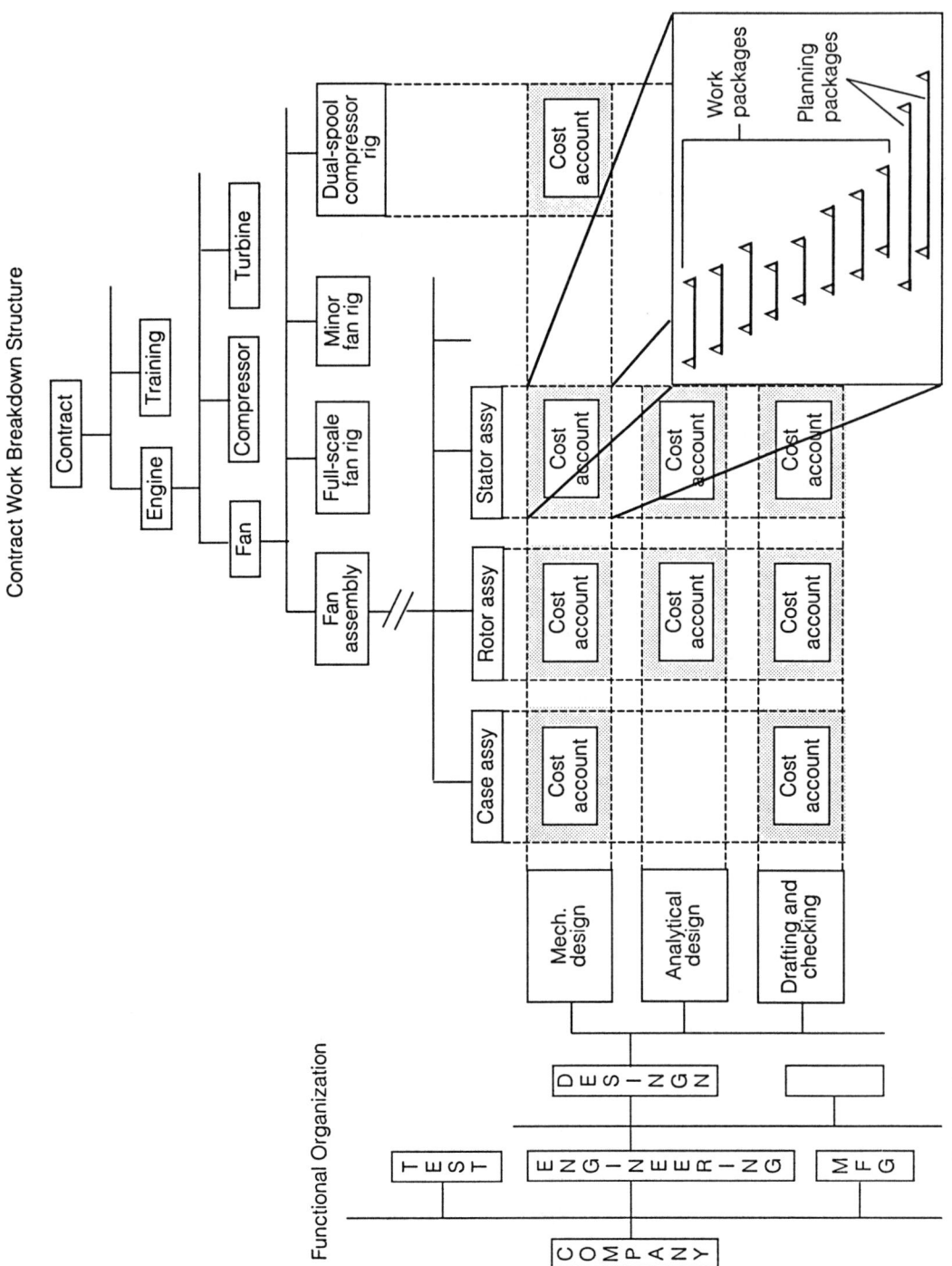

Fig. 17-4. *Integration of WBS and organizational structure.* (Source: Cost/Schedule control systems criteria joint implementation guide)

Work breakdown structure

PROJECT	WORK BREAKDOWN STRUCTURE DICTIONARY (STATEMENT OF CONTENT)	DATE	REV
		SHEET	OF

WBS NUMBER	ELEMENT TITLE

Element Description

Objective:

Included:

    Brief statement of included work.

Not included:

    Work that could be assumed to be included but is not.

Documents reference:

    Reference paragraph number of statement of work, specification, etc.

Associated Lower-Level Elements

WBS NUMBER	TITLE

Fig. 17-5. WBS dictionary example.

The number of levels of a Work Breakdown Structure depends on:

- Project size.
- Work packages size.
- Schedule.
- Implementation cost versus the benefit.

You have completed defining a Work Breakdown Structure element when the smallest subdivision of work:

- Can be assigned to one person for responsibility and requires not more than three people to perform it.
- Can be allocated resources.
- Can be proven it's done when completed.
- Can be assigned precedence in regard to other tasks.

The start and stop points of Work Breakdown Structure elements are not tied to calendar dates. They are later tied to time duration through the work packages. After the Work Breakdown Structure and work package requirements are complete, calendar dates for work packages are calculated based on absolute dates, duration of tasks, and resource availability.

Time phasing the work

It might not be practical or possible to do all detailed planning at the beginning of the project. Instead, large increments of work could be identified and given budgets based on prior experience or gross estimates. As the work is defined, a "rolling wave" planning concept is used. Near-term work is segregated into planning packages. Planning packages are similar to work packages but have gross tasks, schedules, and budgets that are detailed into work packages with smaller tasks. Efforts are progressively divided into smaller subdivisions as work proceeds. Work package definition must happen in sufficient time to complete detailed planning and budgets.

To insist on "a" Work Breakdown Structure for a project is an error. The work to be done during design is different from that of manufacturing or production. Identify natural phases in the project's life cycle and create a Work Breakdown Structure for each phase. Figure 17-6 illustrates the rolling plan concept for definitization of the work effort. Near-term work is finely definitized as each phase approaches. You cannot define the work to be done in product integration until you have defined the product itself, for example.

Development steps

Developing a Work Breakdown Structure follows these 10 steps:

1. Subdivide the total effort into discrete elements.
2. Satisfy deliverables (hardware, software, and services).

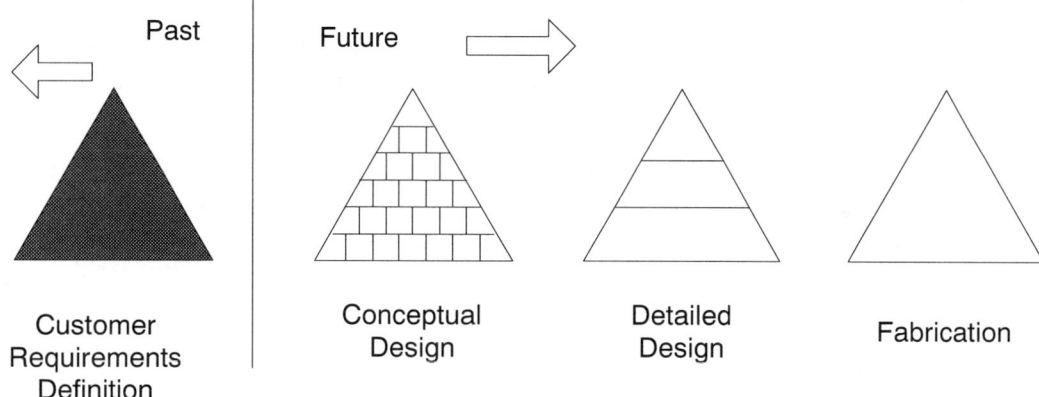

Fig. 17-6. Detail increases near-term.

3. Correlate the elements with the statement of work, the specification, contract line items, data items, and other requirements.
4. Include not only discrete effort, but also apportioned effort and level of effort tasks as appropriate.
5. Make sure the cost of an element is the sum of the costs of the elements feeding into it.
6. Rearrange elements as needed.
7. Make a Work Breakdown Structure dictionary.
8. Design a work package form and instructions.
9. Have the work packages completed.
10. Continue to work with the project team until both structure and work packages satisfy your control system.

You can develop the Work Breakdown Structure in a number of ways. If you have software that supports Work Breakdown Structures or software for outlining, some of the burden is relieved. The process is iterative and automatic numbering and block moves can help you. A Work Breakdown Structure should never be held up waiting for software, however. Pencil and paper are still effective ways of making the Work Breakdown Structure. Post-it note pads, self-adhesive masking paper, or forms taped to a wall can help you organize the work.

Criteria for inclusion

The criteria for selecting work to be included are deliverables and accomplishments. Deliverables are items you produce for your customer or an internal customer. Accomplishments are items necessary to ensure the success of the project.

Deliverables include such items as prime equipment, test equipment ordered by the customer, software programs, data items, manuals, training courses, and field service.

Accomplishments might include providing engineering workstations within the project, implementing an electrostatic discharge control plan, training manufacturing on a new assembly process, or hiring more engineers. Accomplishments also include design, fabrication, test, or delivery.

Types of effort

Efforts fall under one of three categories:

1. *Discrete effort*—tasks that have a specific end product or end result.
2. *Apportioned effort*—an effort that is a direct function of other discrete tasks. That is, it can be calculated by applying a factor to a directly related task.
3. *Level of effort*—work that does not result in a final product, but instead is more time-related. Examples are program or project management, contract administration, and secretaries. It is necessary to hold the level of effort to a minimum and identify it separately from the work package effort.

Identifying work to be done

Some of the initial sources for work to be done are:

- The contract.
- The purchase order.
- The statement of work.
- The request for proposal.
- System requirements.
- Conversations with the customer.
- Conversations with your management.
- Conversations with the project members.
- Experience in related projects.
- Your own judgment.

Developing work packages

Work packages are natural subdivisions of elements and are the basic building blocks for planning, controlling, and measuring work performance. A work package is simply a low-level task or job assignment. It describes the work to be accomplished by a specific person or team.

Evaluating work performance drives the desirability of short-term work packages. When work packages are short, little or no assessment of work-in-process is required and evaluating progress status is possible mainly from work package completions. The longer the work packages, the more difficult and subjective the work-in-process assessment becomes.

As a rule of thumb, you should be able to track progress with a one or two percent resolution. On engineering projects, this means work packages are typically 20 to 80 hours in duration and contain 20 to 240 man-hours of effort. The engineering work package definition can be difficult because of the dynamic nature of unprecedented work. In this case, a rolling forecast and definition for the near-term of one or two months is a possible solution.

Work packages might include:

- Tasks in a brief statement of work.
- Relationship to the Work Breakdown Structure (number).
- Responsible party.
- Resource requirements.
  - ~ Type of people, number, and durations.
  - ~ Materials.
  - ~ Equipment.
  - ~ Facilities.
  - ~ Funds.
- Duration.
- Precedence with other work packages.
- Schedule, if calendar-date critical.
- Cost account numbers.

According to the *Cost/Schedule Control Systems Criteria Joint Implementation Guide*, work package characteristics:

- Represent units of work at levels where work is performed.
- Are clearly distinguishable from other work packages.
- Are assignable to a single organizational element.
- Have scheduled start and completion dates and, as applicable, interim milestones, all of which are representative of physical accomplishment.
- Have a budget or assigned value in terms of dollars, man-hours, or other measurable units.
- Are limited to a relatively short span of time or subdivided by discrete value milestones to facilitate the objective measurement of work performed.
- Are integrated with detailed engineering, manufacturing, or other schedules.

You might also say that effective work packages represent a specific, definable unit of work and identify specific accomplishments (outputs) to result from a unit of work; e.g., reports, hardware deliveries, tests.

Some considerations in defining a work package include the:

- Cost or total man-hours of the work package.
- Calendar duration needed to complete the work.
- Skill and experience of the people doing the work.
- Assignment of single-point responsibility.

Other factors of defining a work package become apparent when you use a computer to manage your work. Most computer programs assume that a work package cannot be divided, and that it must be completed without interruption once started. It is also assumed that all resources (people, money, equipment) allocated to that work package are tied up for the entire duration of the work package.

If a resource is needed for only part of a work package, you might consider further subdividing the work package. For example, suppose you are making 10 unique mechanical piece parts and you want to run a vibration test on each of them individually. You have 10 work packages to design and build sample parts, one for each unique part. Each work package calls for your one and only vibration table. You will have an artificial resource conflict since it is highly unlikely that all the parts will be using a vibration table before their design is complete!

Preparing work packages is a team effort. You must have input from those who will actually do the work. You can best do your job by setting up the communications system. Define common meanings for labor, resources, costs, and schedule. Communicate a standard form to the project members. Hold a meeting after inputs are collected and resolve any conflicts or duplicated work.

Handling system-level efforts

Work to be done at the piece-part level is easy to define as a work package. Work done for modification and changes, such as redesign, rework, retooling, retesting, and refurbishing, stays with the affected part's work package. The cost account will overrun, and you must not use another account number to hide the overrun. This will distort the information fed to the control system.

But what about work not associated with a piece part? What do you do with assembly and integration work at the system level, for example? These efforts are separate work packages. Many systems engineering tasks belong at the top of the Work Breakdown Structure because they affect the whole system, not a part of it. An example is reliability predictions. If the prediction is for an electronic board, the effort belongs with the board's work package. If the reliability prediction is for the system, the effort belongs in a work package for system-level effort.

Integration and assembly efforts include the:

- Development of engineering layouts, the overall design characteristics, and the requirements of design reviews.
- Testing of assembled components or subsystems prior to installation.
- Final assembly of components to form a complete product.
- Conduct of product acceptance testing.

Estimating pitfalls

Engineers are usually optimistic in estimating. Frequently, they leave out time to test, debug, and correct their designs. This time can be as much as the design time. Other resources often forgotten are:

- Specialty engineering.
- Data management.
- Configuration management.
- Drafting.
- Modeling and simulation time.
- Test procedure and test set.

Making revisions

Plans are tools for achieving project success. They should not become "frozen" and unchangeable. Often, the project is changed by new realities in the marketplace or in technology. The work to be done changes in step with changing conditions. When you must make changes, follow these guidelines:

- It is all right to replan unopened work packages within cost budgets.
- Do not reopen closed work packages or make retroactive changes.
- Do not transfer work or budget independently, that is, a budget of resources ties to work to be accomplished.
- Do not change in-process work packages.

Recognize changes in the future. The past cannot be changed. Changes to on-going work packages prevent control and encourages poor planning and budgeting.

Summary

- The Work Breakdown Structure is a tool. Its value is in organizing information for:
    - ~ Critical path
    - ~ Schedule
    - ~ Risk
    - ~ Staff
    - ~ Responsibility
    - ~ Resources
    - ~ Budgets
- A common error in developing a Work Breakdown Structure is to subdivide work according to people-functional organization.
- A Work Breakdown Structure book saves time when developing and using WBS.
- To insist on "a" Work Breakdown Structure for a project is an error. The work to be done during design is different from that of manufacturing and production.

- There are two criteria for selecting work to be included: deliverables and accomplishments.
- Efforts fall under one of three categories:
  - Discrete effort.
  - Apportioned effort.
  - Level of effort.
- Engineering project, work packages are typically 20 to 80 hours in duration and contain from 20 to 240 man-hours of effort.
- Preparing work packages is a team effort. You must have input from those who will actually do the work.
- Engineers are usually optimistic in estimating. Frequently, they leave out time to test, debug, and correct their designs.

Further reading

U.S. Department of Defense. *Cost/Schedule Control Systems Criteria Joint Implementation Guide*. AFSC P173-5, AFCC P173-5, AFLC P173-5, AMC-P 715-5, NAVSO P3627, DLAH 8400.2, or DCAA P7641.47. Washington, D.C.: U.S. Department of Defense, 1987.

18

Cost and schedule control

Cost and schedule control are important to systems engineering because:

- Both cost and schedule are constraints on the systems engineering process.
- Products late to market can miss the market.
- Cost is a measure of limited resources consumed.
- Schedule slips identify places you are stuck, and possible technical problems.
- The schedule communicates the plan, and allows integrated and coordinated engineering efforts.

In this chapter, you'll learn:

- What makes a schedule effective.
- How to look at cost.
- The basics of earned value.
- Where to look for microcomputer tools.

Control techniques

This chapter is not a comprehensive guide to cost and schedule control. You should have a project manager who is competent in these areas. Managing systems engineering does require a certain proficiency in control techniques, however. You usually have a deadline for bringing the product to market and limited resources to do so. You would like to do this as quickly as possible with the minimum required resources. Cost and schedule control are your tools towards achieving these goals.

Cost and schedule tie closely to work breakdown structures and work packages. Costs and time should be easily extractable from work packages for roll-up at higher levels.

Cost and schedule planning and control is done most easily by computer. Microcomputers and software are readily available at low cost. Systems engineers can learn most software in under a week and use a tutorial the first hour.

Systems engineering plans heavily at the beginning of a project or during a change of phase. Revisions are made as needed or to speed up the project under competitive pressure. Therefore, the planning and control system must be flexible and easy to modify.

Defining control

Controlling means helping the engineering staff implement the plan. The engineering staff designs the product to meet the requirements; the plan or schedule does no design work. The plan and schedule are necessary to deal with complexity. But leading the project by deviations from the plan is like driving while always looking in the rearview mirror. It is an important check but it is not sufficient by itself to manage systems engineering. Planning and controlling cost and schedule includes the following tasks:

- Establish and support the project's objectives.
- Define the work to be done.
- Determine the resources needed.
- Schedule the tasks.
- Establish the costs.
- Evaluate and optimize the schedule and resources.
- Establish the cost and schedule plan.
- Compare actuals against planned.
- Analyze, correct, and change plan where needed.

As you establish the plan, you are concurrently defining the work to be done, the resources required to do the work, and the timing of that work.

Schedules

Scheduling includes both the planning and the execution of the plans. A schedule is a time plan of goals or targets that serves as the focal point for management actions. Milestones are events of particular importance. Tasks in the schedule can sometimes be sequential or concurrent. Sequential means serial, or one thing after another. Concurrent means two or more tasks done at the same time or at overlapping times. Scheduling focuses on three things:

1. Tasks.
2. Milestones.
3. Constraints or dependencies.

Constraints or dependencies are things that cannot happen until something else happens first.

Schedule types

Systems engineering uses three types of schedules:

1. Gantt or bar charts.
2. Milestone charts.
3. Network schedules.

Gantt or bar charts focus on planned activities. They are easy to read and most people are familiar with them. They can be used to communicate with work groups or to show short-term work schedules.

Milestone charts focus on events in time. They show the big picture. Results, not the means, are the theme of the milestone chart. Use them to communicate with management and to coordinate several work groups.

Subsets of the network charts include CPM, PERT, and PERT flavors (Project Evaluation and Review Technique). Network charts tie the tasks together through constraints and dependencies. The critical path is the longest sequence of connected tasks through the network, from start to finish. These types of charts deal better with the complexity of systems engineering for scheduling. However, networks require well-defined tasks.

PERT grew out of the challenge of producing the U.S. Navy Polaris missile around 1958. Employees from Lockheed Aircraft Corporation, the Navy Special Projects Office, and Booz, Allen and Hamilton worked on reducing the time element of the project. The result was the PERT tool for planning and scheduling with its emphasis on time. A major feature of the tool was the recognition of probability in task completion and scheduling.

CPM (Critical Path Method) was formulated at about the same time from du Pont and Remington Rand Univac. The objective was to reduce the time required to perform plant construction, overhaul, and maintenance. Simply, they were interested in trading time and project cost. Because their tasks were well-defined, CPM is less focused on time and emphasizes minimum project costs. Neither the PERT nor the CPM group was aware of the other until 1959, hence the two different techniques.

Precedence diagrams were introduced later. The focus is on tasks that are linked by dependency lines. This technique is most popular today because it allows greater flexibility in replanning.

You might find resistance from older managers who remember being forced to use PERT/CPM before the days of microcomputers. Tales of paper printouts wrapping around halls, and waiting weeks to make changes are real! This is no longer the case when you can do these control techniques at your desk, however. In fact, you don't need to understand the mathematics because project management software will do that for you. PERT and CPM were probably 25 years ahead of the computer revolution needed to make them accessible.

The best advice for immediate results is to find and use a good software package. The packages come with input screens and graphical outputs. Outputs are available as

Gantt charts or milestone charts for presentation purposes. Detailed knowledge of the mathematics is not necessary. If you wish to know the details, the techniques are well documented under the topics of PERT, CPM, and precedence diagraming.

Scheduling effectively

Effective schedules:

- Are easily understandable.
- Identify critical tasks (such as time critical).
- Can provide details for evaluating resource use.
- Are compatible with the master plan.
- Are easily modified for changes.

Measuring progress by schedule

Progress should not be measured by "percent complete." The emphasis belongs on the time remaining to complete the tasks. How much time is on the critical path? Progress measurement follows three steps:

1. Record progress-to-date by finished tasks.
2. Estimate remaining work.
3. Determine the impact to interface, deliverables, and milestones.

Progress must be measured at least weekly. If the schedule is tight, measure daily or every other day. Identify problems quickly and decide what to do about them.

Watch the critical path. On a typical project, about 15 percent of the tasks are on the critical path. The Pareto principle states that 80 percent of your problems will come from 20 percent of the tasks. The "significant few" are probably on your critical path. There are three usual ways to measure task progress:

1. Show 0 percent progress until task completion, then 100 percent.
2. Show 50 percent on start, 100 percent at completion.
3. Estimate a continual percent progress until completion.

The first method is preferable. It tends to underestimate completion but, if the tasks are small, this is not much of an error. The third method leads to the "90 percent complete" syndrome. The progress estimator tends to be the task performer. Wishing to show progress, the performer will raise the percentage at each review period. Progress rapidly rises to 90 percent complete, where it stays for long periods of time. When the performer runs into difficulties, the percentage is never lowered to show this. You soon have documentation showing the project 90 percent complete, which, in fact, is a fairy tale. Keep the tasks small, and only give credit for full completion against exit criteria.

Another fairy tale is measuring progress by the funds expended. Money spent has nothing to do with progress. Many projects have spent a great amount of money only to find themselves farther from the goal than when they started. Don't measure progress by money spent.

Dealing with complexity

Deal with complexity by breaking down schedules. The master schedule is usually a milestone chart with management-mandated events. Examples of schedules that must be compatible with the master are:

- Development
- Integration
- Test and verification
- Production

Time phasing is another way to deal with complexity. Use Gantt or milestone charts for future events that are fluid. Use more precise scheduling for near-term events and tasks. Spread the burden of scheduling through the project life.

Defining task timing

Assuming you have a master schedule from the project manager and have prepared work packages, how do you schedule the tasks? For systems engineering, you must satisfy two constraints, usually recursively.

The first constraint deals with time. Each work package should have a time duration associated with the task. Most tasks are not calendar-driven, they are free to start when you want. Other tasks can be tied to a calendar date, such as a new component from a supplier or a marketing promotion date. You must establish the precedence and dependencies for all tasks. When all tasks are linked, you have a schedule that assumes no other constraints. But there is constraint number two.

Constraint number two deals with the available resources. Resources include cash flow, equipment, facilities, and people. You are limited by resources to compress your schedule. Two people cannot use the same vibration table. You can't schedule your only safety engineer to work three person-months in April. These conflicts must be resolved. Resource-leveling refers to evening out the use of resources. For example, you prefer not to have overtime nor people sitting around with no tasks scheduled for them to do. A computer allows the recursive juggling and "what if" trials needed to optimize the schedule.

Crashing a schedule means to shorten it. Always begin on the critical path. Shorten the activity that gives up the most time for the least cost and risk. You might have to add resources or buy the expected result from another source. Adding people does not always speed things up. In fact, it can add time to the schedule. They must be the right people to add with the right skills and behaviors you need for that task. If too many new people are added, the time spent communicating with them slows down the task. Be very careful. Adding 50 people to dig a fence post hole results in one person digging and 49 people bothering the person with questions and advice.

Sometimes, shortening the critical path will cause a new critical path to appear. This also frequently happens when tasks on "near-critical paths" are delayed, causing a new critical path. You should pay attention not only to the critical path, but also to near-critical path tasks.

Systems engineering's affect on cost control

You might wonder why cost control is a topic for systems engineering. Isn't that the project manager's job? It is a job for both functions because costs are really measures of resources used. Systems engineering influences cost through three ways:

1. Technical approach.
2. Risk.
3. Scheduling of tasks and resources.

In turn, budget and cash flow are constraints on:

- Systems engineering resources.
- Technical approach.
- Time to complete the project.

Yes, systems engineering does drive costs and is concerned when budgets change. Costs exceeding budgets are signs of trouble.

Cost control at the project level consists of:

- Preparing budgets down to specific tasks.
- Verifying that charged costs are right.
- Comparing actual costs to budgeted costs.
- Acting on and controlling variances as appropriate.

The project requires cost reports weekly for timely control. Waiting a month for costs allows no time for corrective action but only records historical problems.

Costs can be summarized in three ways by:

1. Elements of the work breakdown structure.
2. Financial ledger, such as labor, materials, overhead, etc.
3. People organization, such as departments, cost centers, etc.

Reports within these summaries include:

- Actual costs versus budgeted costs.
- Variances.
- Labor time and wages.
- Purchase order summaries.
- Open commitments (orders placed but not received).

The budget establishes the cost baseline, and the people affected by the budget should participate in budgeting decisions. Lack of involvement in setting objectives often leads to less effective efforts. Stress is reduced by having a say, and the participants have more of a sense of ownership. As a practical matter, work packages must have the input of those doing the work, as they are usually most qualified to estimate the work required on engineering efforts.

The cost baselines must be carefully structured to allow cost tracking. It should be easy to pull both people organization costs and work breakdown package costs from the collection system. You want to know how much it really costs to do the work package. Not only is this important to track use of resources, but the next time you estimate, you would like to do better at it.

Earned value

As stated earlier, money spent is not a measure of progress. It is important to track both cost and schedule deviation, however. Comparing actuals against estimates enables you to forecast final costs and completion date.

Progress must be tracked both at the project level and at the work package level. The project level gives a trend for the total effort. Some work packages will be ahead and some behind. Even if the project looks good, always check at the work package level to see what is happening.

Analyzing every work package is unnecessary and wasteful of your time. Establish cost and schedule variance thresholds and analyze only those variances that are significant. Thresholds should have both percentage and fixed amounts. This will prevent small work packages from being flagged for large percentage variance, but relatively small fixed amounts.

Understanding cost and schedule reports

Five terms that can help you to understand cost and schedule reports are:

1. **BCWS**—Budgeted Cost for Work Scheduled. The plan against which the performance is measured. The total budgets for all work packages, apportioned effort, and level of efforts.
2. **BCWP**—Budgeted Cost for Work Performed. The earned value of the work accomplished against the plan. The sum of the budgets for completed work packages, and the applicable portions of budgets for level of effort and apportioned efforts.
3. **ACWP**—Actual Cost of Work Performed. The cost actuals. The costs actually incurred and recorded in accomplishing the work performed within a given time period.
4. **BAC**—Budget at Completion. Sum of the budgeted plans at the end of the project.
5. **EAC**—Estimate at Completion. Latest forecasted cost at the end of the project.

Figure 18-1 displays the cost and schedule status against the plan. The schedule is projected to slip beyond the original completion date. The project is forecast to overrun by the excess of the Estimate at Completion (EAC) over the Budget at Completion (BAC). The original Budget at Completion was the target cost for the project.

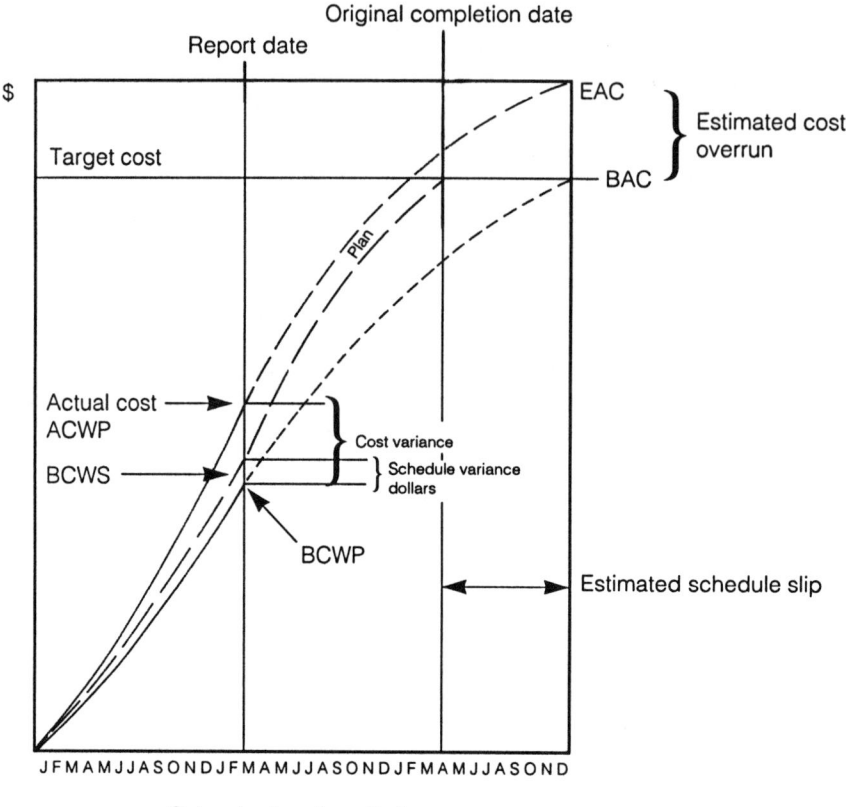

Fig. 18-1. *Cumulative plan/status display.*

The report date has three costs associated with it. They are the actual cost to date, the Budgeted Cost for Work Scheduled (BCWS), and the Budgeted Cost for Work Performed (BCWP). The cumulative actual costs are higher than the planned costs, or the BCWS. The earned value, or BCWP, is less than the plan. This project is spending money faster than the plan and yet is behind the plan in actual work accomplished.

The difference of the actual costs and the BCWS is the cost variance at this reporting date. The difference of the BCWS and the BCWP is the schedule variance in dollars. The schedule variance in time can be estimated by looking at the BCWS and moving across to the right to intersect the projected BCWP line. The end of the project shows several months of schedule slip.

If the project is more than 15 percent complete, the trends are established. Overruns cannot be made up; they are more likely to grow to greater percentages. If the near-term is underestimated, the estimates on the far-term are probably worse. This is not taking a negative view, it is the experience of more than 200 major projects. If the project is behind, admit it; for you will not catch up.

Estimating cost and time at completion

A simple formula to estimate the cost at completion is:

$$\frac{\text{ACWP}}{\text{BCWP}} \times \text{BAC}$$

To estimate the time at completion:

$$\frac{\text{BCWS}}{\text{BCWP}} \times \text{original completion date}$$

Table 18-1. Sample Cost/Schedule Diagnosis.

If	Then consider:
Under budget	Behind schedule with tasks? Properly staffed? Suppliers charges recognized? Breakthrough in work efficiency? Long lead or critical purchases made?
Over budget	Technical difficulties? Erroneous charges? Overstaffed? Estimate too optimistic? Need to re-forecast project? Purchases made too early? Change in scope without approval and allocation of funds for changes? Clear objectives?
On budget	Scheduled work packages completed?
Behind schedule	Look at critical path Risk management Assess resources; right people to do tasks?
Ahead of schedule	Clear objectives? Spending unnecessarily to expedite? Have task exit criteria been met?
On schedule	Have task exit criteria been met? Holding walk-throughs, reviews, and audits to verify meeting requirements?

Microcomputer tools

The microcomputer is a powerful tool for determining project cost and schedule control. Hardware and software are affordable, especially when compared with the benefits received. Also, the ability to do "what if's" allows better optimization of project control. There are many good software packages for project management, and the latest information and comparisons can usually be found in: *Industrial Engineering, PC Magazine, PC World, Software Digest Ratings Report,* and *Survey of Project Management Software Packages*, Project Management Institute, P.O. Box 43, Drexel Hill, PA 19026.

Summary

- Cost and schedule control are important because:
  - Both cost and schedule are constraints on the systems engineering process.
  - Products late to market can miss the market altogether.
  - Cost is a measure of limited resources consumed.
- Controlling means helping the engineering staff to implement the plans.
- As you establish a plan, you are concurrently defining the work to be done, the resources required to do the work, and the timing of that work.
- Scheduling focuses on three things:
  - Tasks.
  - Milestones.
  - Constraints or dependencies.
- Effective schedules:
  - Are easily understandable.
  - Identify critical tasks (such as time critical).
  - Can provide details for evaluating resource use.
  - Are compatible with the master plan.
  - Are easily modified.
- Progress should not be measured by "percent complete." Instead, the emphasis belongs on the time remaining to complete the tasks.
- Money spent has nothing to do with progress.
- Crashing a schedule means shortening it. Always begin on the critical path. Shorten the activity that gives up the most time for the least cost and risk.
- Systems engineering influences cost through:
  - Technical approach.
  - Risk.
  - Scheduling of tasks and resources.
- Establish cost and schedule variance thresholds and analyze only those variances that are significant. Thresholds should have both percentage and fixed amounts.
- If the project is more than 15 percent complete, the trends are established. Overruns cannot be made up; they are more likely to grow to greater percentages.

Further reading

Archibald, Russell D. *Managing High-Technology Programs and Projects*. New York: John Wiley & Sons, 1976.

Badiru, Adedeji B. and Gary E. Whitehouse. *Computer Tools, Models, and Techniques for Project Management*. Blue Ridge Summit, PA: TAB Books, 1989.

Cleland, David I. and William R. King, eds. *Project Management Handbook*. 2nd Ed. New York: Van Nostrand Reinhold, 1988.

Moder, Joseph J., Cecil R. Phillips, and Edward W. Davis. *Project Management With CPM, PERT, and Precedence Diagramming*. 3rd Ed. New York: Van Nostrand Reinhold, 1983.

Thamhain, Hans J. *Engineering Program Management*. New York: John Wiley & Sons, 1984.

19

The system engineering management plan

The System Engineering Management Plan is important because it:

- Sets technical objectives and goals.
- Defines tasks and responsibilities.
- Provides a basis for directing, measuring, and controlling progress.
- Plots the path to optimum results through limited resources, capabilities, and changing conditions.

In this chapter, you'll learn:

- The objectives of a System Engineering Management Plan.
- Planning basics.
- How to use a System Engineering Management Plan.
- How to apply MIL-STD-499A, *Engineering Management*.

Overview

The System Engineering Management Plan (SEMP) is a concise, top-level management plan for integrating all of the system activities. It makes visible the organization, direction, and control mechanisms and people for the attainment of cost, performance, and schedule objectives. It defines the who, what, when, where, how, and why of the decision-making process, and is as detailed as appropriate to the system's life cycle phase and degree of system complexity. It should reflect good management judgment with minimum documentation.

The major objectives of the System Engineering Management Plan are to:

- Facilitate communications.
- Integrate all engineering disciplines.

- Ensure the product meets the requirements.
- Establish streamlined checks and controls.
- Define the system engineering process.

The SEMP defines and describes the type and degree of system engineering management, the system engineering process, and the integration of engineering efforts. The plan identifies:

- Organizational responsibilities.
- Authority for system engineering management.
- Levels of control for performance and design requirements.
- Control methods to be used.
- Technical program assurance methods.
- Control procedures to ensure integration of requirements and constraints.
- Schedules for design and technical program reviews.
- A detailed description of the system engineering process to be used.
- Specific tailoring to requirements of the system in-house documentation.
- Trade-off study methodology.
- Types of mathematical and simulation models to be used for system and cost-effectiveness evaluations.

Depending on the system peculiarities, the plan should also describe the special or intensive management aspects of functions and activities critical to the system objectives. These might include, for example, risk analysis and assessment, resource allocation, work elements, trade-offs, program assurance, and many other specialties. As a top management tool, the SEMP must present the system engineering management and system engineering process, and relate these to the engineering specialties, activities, and functions as an integrated plan, rather than as a composite or summary of discrete subplans.

Planning in general

Planning *is* decision-making. A written plan is a documented set of decisions. Four elements in general planning are the:

1. Work to be done.
2. Timing.
3. Resources.
4. People.

The general components of management plans are:

- Objectives
- Deliverables
- Schedule/milestones
- Approach

- Resources
- Organization/responsibilities
- Interrelationships
- Staffing
- Procedures
- Planning and control
- Progress reporting
- Exit criteria
- Training
- Risk management

Using the plan

The System Engineering Management Plan:

- Is tailored.
- Is a living document.
- Dictates minimum documentation.
- Applies resources cost effectively.
- Is prepared by the system engineer.
- Is compatible with other project plans.

Tailoring the SEMP

The SEMP must be tailored to both breadth and depth. Tailoring in breadth includes eliminating or adding system elements, particular system engineering activities, and functional areas. Tailoring in depth involves deciding on the level of detail required to identify, describe, and specify the requirements. The depth of system engineering varies for each project in relationship to its complexity, uncertainty, urgency, and risk threshold.

Planning can be overdone. Don't dismiss process improvements because "they're not in the plan." Blind adherence to past planning ignores newly identified risks, and increasing knowledge. Processes that are not working need attention. You should identify simplification opportunities and rank them for all processes.

Time phase the SEMP

The System Engineering Management Plan is a time-phased document that should be continuously updated—a living plan. Do not prepare a SEMP, and discard it into the bottom drawer of a filing cabinet. It is not a data item to be checked off, it is your tool for managing the system engineering effort and technical effort of your project. Make the System Engineering Management Plan modular to break down complexity. Establish time periods or criteria to update the document to break up time.

Constraining documentation

System and design engineering are recursive activities. Ask for minimum documentation to prevent engineering bottlenecks. Documentation is not a substitute for effective systems engineering. Don't saddle design engineers with documentation that only vaguely adds value to the product. Technical decisions are usually made directly, with only a fraction of data used as a deciding factor. Make sure documentation adds value to the end product.

Applying resources cost effectively

On almost any project, there are technical areas with varying degrees of potential performance and cost benefits. Limited resources should be expended where the payoff potential is the highest. Areas that include the possibility of either increasing performance capability or reducing operating and maintenance cost.

In applying system engineering to a specific project, a major consideration should be the benefits expected versus their cost. This applies both to the types of procedures to be used and the depth of detail to which the process is carried out and managed. Neither the rigor nor the depth of the procedure used should be greater than their worth to the project. The relationship between expenditure of engineering analysis time or testing and level of confidence is usually nonlinear. Sometimes, the potential value of increased confidence beyond a certain level does not justify the added cost.

The system engineer judges where efforts are most needed. The system engineer must write the System Engineering Management Plan, it cannot be given to a junior engineer. Based on the system engineer's experience, he or she assesses risks and makes trade-offs. These decisions are the responsibility of the system engineer.

Relating to other plans

The system engineering effort must conform with the total project effort. Top project documents include the:

- Statement of Work.
- Milestone schedule.
- Program Plan.
- Work Breakdown Structure.

Other plans that must be compatible with the SEMP include:

- All specialty engineering plans.
- Manufacturing.
- Capital equipment.
- Tooling.
- Verification.
- Budget.
- Marketing.
- Facilities.
- Configuration Management.
- Product support.

Measuring SEMP effectiveness

If you are writing or evaluating a System Engineering Management Plan, how do you measure effectiveness? Goodness is difficult to measure. At the end of the project, you are never sure if success was because of planning, execution, or both. A wonderful plan allows for less than perfect execution but minimizes waste. Here are some sample questions you might ask yourself:

- Are top requirements, results, and deliverables identified?
- Are progress and performance measures stated?
- Are meetings, reviews, and status reports planned?
- Are supplier controls established?
- Are task leaders identified?
- Are technical staff compatible with tasks?
- Are people organization-appropriate?
- Are resources and timing identified?
- Are resources compatible with tasks?
- Is information flow planned?
- Is the technical system robust against external influences (noise)?
- Are controls adequate and at right points?
- Is paperwork simplified?
- Is the plan realistic? Effective? Integrated? Tailored?

If you are writing a system engineering management plan, one of the indicators of effectiveness is your own feelings. Feelings of uncertainty and stress that the project is running you, or that you can't fill out the plan, are warning feelings. You are probably in over your head and should seek help from another competent system engineer.

Improving the system engineering process

The system engineering process is made up of many smaller processes. When the tasks and activities have inputs, outputs, standards, procedures, and time references, they become processes. Process improvement and cycle time reduction go together. A summary of improvement for processes is to:

- Establish ownership and responsibilities of the process.
- Define the process and identify the requirements and expectations.
- Define and establish measures for the process.
- Assess conformance with requirements and expectations.
- Identify improvement opportunities—cycle time reduction or elimination of non-value added efforts.
- Rank improvement opportunities and set targets for improvement.
- Improve the process.

Applying MIL-STD-499A

A System Engineering Management Plan (SEMP) is created to structure engineering planning, processes, and outputs. Engineering Management for the Department of Defense is described in MIL-STD-499A. This document provides standard criteria for evaluating engineering planning and output. MIL-STD-499A divides systems engineering management into three types of activities:

1. Technical program planning and control.
2. System engineering process.
3. Engineering specialty integration.

The structure of MIL-STD-499A activities are shown in FIG. 19-1. You can find the entire document of MIL-STD-499A in Appendix A of this book. MIL-STD-499A defines Engineering Management as:

> The management of the engineering and technical effort required to transform a military requirement into an operational system. It includes the system engineering required to define the system performance parameters and preferred system configuration to satisfy the requirement, the planning and control of technical program tasks, integration of the engineering specialties, and the management of a totally integrated effort of design engineering, specialty engineering, test engineering, logistics engineering, and production engineering to meet cost, technical performance and schedule objectives.

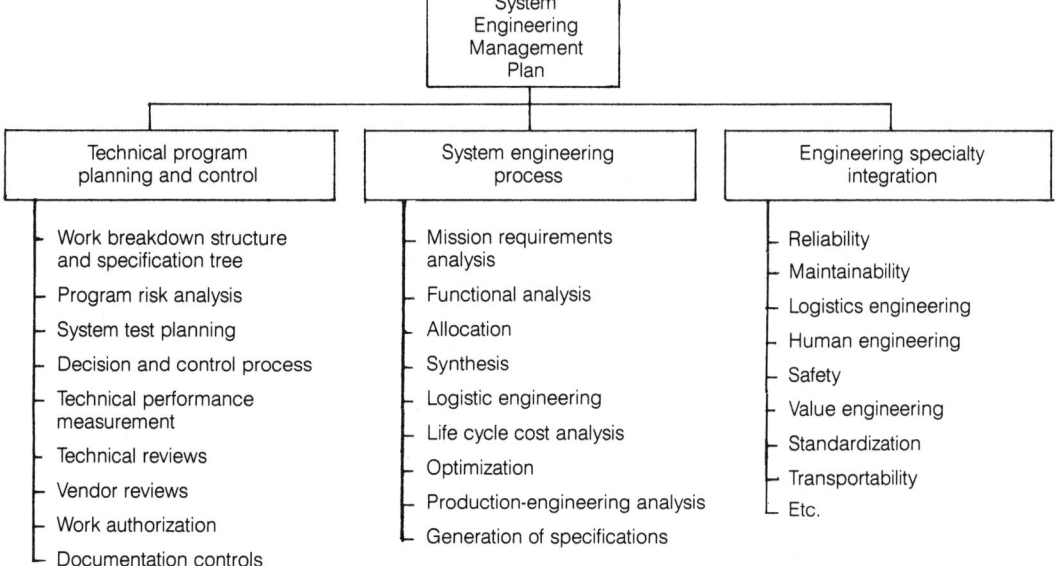

Fig. 19-1. MIL-STD-499A suggested tasks.

MIL-STD-499A is clear about the responsibility of the technical effort. Technical performance of the product is only one piece of the total effort. The result must meet cost, schedule, and technical performance objectives. An engineering project measurement includes cost, schedule, and technical performance.

You might also note that managing the engineering effort includes testing, logistics, and production engineering. Design is not the only, nor the end product, of the engineering team.

Technical program planning and control

The first major category of engineering management in MIL-STD-499A is Technical Program Planning and Control. MIL-STD-499A defines this as:

> The management of those design, development, test, and evaluation tasks required to progress from an operational need to the deployment and operation of the system by the user.

Technical program planning and control includes:

- Identifying organizational responsibilities and authority for system engineering management.
- Establishing methods and levels of control for performance and design requirements.
- Establishing technical project-assurance methods.
- Planning and scheduling design and technical project reviews.
- Controlling documentation.

System engineering process

The second major area addressed by MIL-STD-499A is the System Engineering Process. Again, referring to FIG. 19-1, activities that are customarily associated with analytical systems engineering also apply to the System Engineering Process. MIL-STD-499A defines the System Engineering Process as:

> A logical sequence of activities and decisions transforming an operational need into a description of system performance parameters and a preferred system configuration.

Planning for the system engineering process includes:

- Describing in detail the process to be used, including specific tailoring to the requirements of the product and project.
- The procedures to be used in implementing the process.
- Documentation.
- Methods for trade studies.
- Types of models to be used for product and cost-effectiveness evaluations.
- Generation of specifications.

Engineering specialty integration

The third major area of concern of MIL-STD-499A is Engineering Specialty Integration. The coordination of specialists and engineering disciplines is not an easy job. Many trade-offs must be made to optimize the project design, and a written plan helps to communicate the actions needed from all members of the project team. MIL-STD-499A defines Engineering Specialty Integration as:

> The timely and appropriate intermeshing of engineering efforts and disciplines such as reliability, maintainability, logistics engineering, human factors, safety, value engineering, standardization, transportability, etc., to ensure their influence on system design.

Engineering specialty integration includes:

- A summary or reference to detailed engineering specialty project plans.
- A description of the integration and coordination of engineering specialties.
- The use of engineering specialty efforts in the system engineering process.
- The interrelations between overlapping specialty efforts.

TABLE 19-1 shows some of the topics addressed in a System Engineering Management Plan. The topics are different for each product and project and the breadth and depth of each topic must conform to each project's particular needs. TABLE 19-2 lists the general criteria found in MIL-STD-499A for evaluating engineering planning and output. Definitions for these terms can be found in section four of MIL-STD-499A.

Tailoring

MIL-STD-499A states that its application must be tailored. Consider both breadth and depth for your particular project when tailoring the system engineering management plan.

MIL-STD-499A will be replaced with DOD-STD-499. Revision A was issued in 1974. Also, the IEEE might publish a commercial standard. You should watch for new developments in standards for systems engineering.

Summary

- The System Engineering Management Plan (SEMP) is a concise, top-level management plan for the integration of all system activities.
- The SEMP defines and describes:
  - The type and degree of system engineering management.
  - The system engineering process.
  - The integration of engineering efforts.
- The major objectives of the System Engineering Management Plan are to:
  - Facilitate communications.
  - Integrate all engineering disciplines.

Table 19-1. SEMP Outline Example.

```
1.0   Technical Program Planning and Control
      1.1    Project Technical Objectives
      1.2    Major Deliverables and Results
      1.3    Synopsis of Major Tasks
      1.4    Milestones/Schedule
             1.4.1   Development Schedule
             1.4.2   Integration Schedule
             1.4.3   Verification Schedule
             1.4.4   Production Schedule
      1.5    Constraints
      1.6    Technical Approach
      1.7    Technical Organization
      1.8    Responsibilities and Authority
      1.9    Technical Staffing
             1.9.1   Key Personnel
             1.9.2   Required Behaviors and Timing
             1.9.3   Workforce Loading
      1.10   Training
      1.11   Standards and Procedures
             1.11.1  National and Regulatory
             1.11.2  Company
             1.11.3  Project Specific
      1.12   Risk Management
             1.12.1  Risk Identification
             1.12.2  Risk Alleviation
             1.12.3  Integration of Risk into Project Plans
             1.12.4  Contingency Planning Procedures
      1.13   Resource Allocation
             1.13.1  Resource Requirements Identification
             1.13.2  Procedures for Resource Control
             1.13.3  Redefinition Procedures
             1.13.4  Reallocation Procedures
             1.13.5  Resource Application
                     1.13.5.1  Equipment
                     1.13.5.2  Facilities
                     1.13.5.3  Funds
                     1.13.5.4  Technical Tools
      1.14   Work Breakdown Structure
             1.14.1  Work Definition
             1.14.2  WBS Setup
             1.14.3  Work Packages
      1.15   Work Authorization
      1.16   Reviews, Walk-throughs, and Audits
             1.16.1  Program Reviews
             1.16.2  Technical Reviews
             1.16.3  Supplier Reviews
             1.16.4  Walk-throughs
             1.16.5  Audits
      1.17   Progress Assessment
             1.17.1  Measures
                     1.17.1.1  Technical Performance Measurements
                     1.17.1.2  Cost
                     1.17.1.3  Scheduled
                     1.17.1.4  Earned Value
             1.17.2  Evaluation
                     1.17.2.1  Corrective Action
```

Table 19-1. Continued.

                  1.17.2.2   Redirection/Redefinition
- 1.18 Change Control Procedures
  - 1.18.1 Baseline Identification
  - 1.18.2 Baseline Change Control
- 1.19 Documentation Controls
  - 1.19.1 Internal Documents
  - 1.19.2 External Documents
- 1.20 Requirements Traceability
- 1.21 Interface Control
- 1.22 Verification Planning
- 1.23 Supplier Control
- 1.24 Standardization, Preferred Parts
- 1.25 Key Technology
- 1.26 Long-Lead Items
- 1.27 Process Improvement
- 1.28 Other Plans and Control

2.0 System Engineering Process
- 2.1 Time Phasing
- 2.2 Mission and Requirements Analysis
- 2.3 Value System
- 2.4 Functional Analysis
- 2.5 Decision Process
- 2.6 Requirements Flowdown
- 2.7 Trade Studies
- 2.8 Design Optimization
- 2.9 Synthesis
- 2.10 Specification Tree/Specifications
- 2.11 Documentation
- 2.12 Verification Process
- 2.13 System Engineering Tools
  - 2.13.1 Quality Function Deployment
  - 2.13.2 Taguchi Methods
  - 2.13.3 Value Engineering
  - 2.13.4 Models and Simulations
  - 2.13.5 Computer-Aided System Engineering

3.0 Engineering Specialty Integration
- 3.1 Integration Design/Plans
  - 3.1.1 Reliability
  - 3.1.2 Maintainability
  - 3.1.3 Supportability
  - 3.1.4 Life Cycle Cost
  - 3.1.5 Human Engineering
  - 3.1.6 Safety
  - 3.1.7 Electromagnetic Compatibility
  - 3.1.8 Testability
  - 3.1.9 Software
  - 3.1.10 Producibility
  - 3.1.11 Design to Cost
  - 3.1.12 Standardization
  - 3.1.13 Other Engineering Specialties
- 3.2 System Integration Plan
- 3.3 Compatibility with Supporting Activities
  - 3.3.1 System Cost Effectiveness
  - 3.3.2 Quality Assurance
  - 3.3.3 Materials and Processes
  - 3.3.4 Manufacturing

242 Management of systems engineering

Table 19-2. General Criteria for Engineering Management (MIL-STD-499A).

(a)	Technical objectives
(b)	Baselines
(c)	Technology
(d)	Realistic system values
(e)	Design simplicity
(f)	Design completeness
(g)	Documentation
(h)	Engineering decision studies
(i)	Cost estimates
(j)	Technical task and work breakdown structure compatibility
(k)	Consistency and correlation of requirements
(l)	Technical performance measurement
(m)	Interface design compatibility
(n)	Engineering specialty integration
(o)	Engineering decision traceability
(p)	Historical data
(q)	Responsiveness to change
(r)	Compatibility with related activities

- ~ Ensure the product meets the requirements.
- ~ Establish streamlined checks and controls.
- ~ Define the system engineering process.
- Planning is decision-making. A written plan is a documented set of decisions.
- The System Engineering Management Plan is a time-phased document. Update the document to make it a living plan.
- System and design engineering are recursive activities. Ask for minimum documentation to prevent engineering bottlenecks. Documentation is not a substitute for effective systems engineering.
- The system engineer must write the System Engineering Management Plan. He or she assesses risks and makes trade-offs. These decisions are the responsibility of the system engineer.
- Process improvement and cycle-time reduction go together.
- MIL-STD-499A states that its application must be tailored. Consider both breadth and depth for your particular project when tailoring the system engineering management plan.

Further reading

U.S. Department of Defense. *Engineering Management*. Military Standard 499A. Washington, D.C.: U.S. Department of Defense.

(Also see appendix A.)

20

Proposals

As a systems engineer, you will both write and read many proposals. Writing technical proposals support sales to customers, and without customers, there is no business. You will also read proposals from your potential suppliers. Selecting the best value is important to systems engineering and proposals are an important communications channel for achieving this. In this chapter, you'll learn about proposal:

- Content
- Planning
- Writing
- Evaluating

This chapter does not attempt to show you technical writing skills, how to manage proposal efforts, or proposal strategy. Each of these topics have many books written about them. Instead, the focus is on writing or evaluating only a technical section.

Training

If you find yourself in charge of a proposal or its strategy, take training before you start. A lot of time and money can be wasted by not knowing what you are doing. There are defined processes for putting proposals together. Proposals don't have to mean heavy overtime and confusion.

Proposals tie systems engineering to sales or purchasing, depending on whether you are writing or reading them. In earlier chapters, manufacturing was tied to systems engineering. The point being made is that systems engineering is integrated with all activities of the business.

Some customers are now asking directly for proof of effective systems engineering, and a proposal is the place to do this. This book is only a baseline on systems engineering from which to start. You, as the systems engineer, must tailor your systems engineering process to your specific organization and product.

Writing proposals

Your objective for the technical proposal is selling the evaluator. A proposal is a sales tool. The evaluator must feel, after reading the proposal, that you understand the problem and have a solution that is acceptable. Therefore, both logic and feelings must be involved. To increase your chances of success:

- Prepare before writing the proposal.
- Follow a plan.
- Ask for, and receive, feedback as you go.

Characteristics of a good proposal include:

- Affordable and acceptable solution cost.
- Addressing concerns and biases learned from direct contact with the customer.
- Substantiated claims.
- Responsiveness to customer requirements.
- Good writing.

Content

Your technical section must show your strong points as benefits and minimize your weak points. Selling your product's features means describing them in the customer's eyes as benefits. For example, reliability means less downtime, lower repair costs, and better dependability on the job. Customers don't buy features or technology, they buy benefits to them. Infer how your product's features are to their benefit.

Proposals solve unresolved problems for the customer. If there were no questions, the customer would simply write a purchase order. Proposals help the customer to make a decision by solving problems for them. Systems engineering is uniquely qualified to do this. To lead the customer through the alternatives and give them a clear choice. Make the decision easy.

Problems must be addressed in three parts. First, explain the problem from the customer's viewpoint. Don't parrot the request for proposal, if you have one. Tell why the customer needs to have this problem solved. Second, show understanding by explaining the difficulties. Demonstrate understanding through your analysis. Third, give the solution through your technical approach. Use the systems engineering tasks to sell your approach over other alternatives. Point out the benefits to the customer.

If you are responding to a request for proposal, make sure you respond to all requirements fully. Follow the instructions exactly. Write to the format asked for, and not to your own.

Your evaluator can consider you deficient in several ways:

- You are nonresponsive.
- You don't meet the minimum requirements.
- You omit facts, use ambiguous wording, or fail to substantiate your claims.
- You propose a high-risk solution without proving its feasibility.

Engineering and business people are usually conservative, nonrisk-takers, and a high-risk solution can be a hard sell.

Your technical section will probably support an overall proposal theme. The theme tells the evaluator your sales message. The theme of your section must also support the rest of the proposal. Themes tie features to benefits. They telegraph the point of the section. Your theme answers the question, "What makes my product beneficial or desirable to the customer?"

Planning your section

Before writing your section, you must first:

- Look at the proposal plan.
- Identify and understand your tasks.
- Analyze the requirements.
- Set objectives.
- Make a compliance checklist.
- Schedule your time to the master schedule.

If you have a request for proposal, look for clues in the:

- Technical description.
- Specification.
- Statement of work.
- Evaluation criteria.
- Schedule.
- Data requirements list.

Using these clues, analyze what the customer is asking. Use a highlighter pen to identify the proposal drivers. What do they really mean by "synthesis" or "systems engineering approach?" What are they trying to tell you?

Writing for evaluators

It helps to know your evaluators. What is their field of expertise? What are their biases? Is there a hidden agenda? Will they have to justify their findings to someone else? Have they already recommended a course of action? Will your analysis make them look bad to their boss? Accurately predicting your evaluator's reaction is an advantage.

Outline your section. Make sure you have covered the issues important to the evaluator. Check that all evaluation criteria are addressed. Discuss the outline with the proposal manager.

Proposals are sales presentations. They must persuade without coming across as a sales pitch. Low-key and credible are features of an effective proposal. Credibility is built by quantifying and detailing. Write positive, specific, and to the point. Make the proposals' message clear:

You must take the viewpoint of the evaluator. What questions could be asked? How much does the evaluator know about the subject? Does the evaluator have a preconceived view? Why will this be a logical solution to the evaluator?

Tell the evaluator why your features are important. Don't assume the evaluator will infer the benefits. Make the evaluator want to read on. Don't ever leave the evaluator asking, "So what?"

Always discuss the benefits before describing how the product is made. Relate each part of the product to objectives of the customer.

Build your technical argument by supporting the customer's intent. Examine the possible alternatives, and tell why the proposed solution is best for the customer. Use the systems engineering approach.

Substantiate your claims. Give proof through sharing your experience or the experience of others. Use photos or models to demonstrate the concepts. Include letters from other experts or customers to support your case. List similar successes. Let the customer decide you know what you're talking about.

Use photos or graphics when words alone won't do the job. In fact, sometimes a photo or video conveys more convincing information than any words. Use both as needed. Convey only one main idea per graphic; it's much more effective.

Be prepared for resistance if your product is an invention or is innovative. Write for credibility. Reduce the perceived risk that goes with newness.

Evaluating proposals

Apply systems engineering tasks as you evaluate proposals. Define the problem well enough so you'll recognize an alternative. Have a value system for later decision-making. Apply systems analysis in choosing the solution.

Create a value system

The value system has the "must have's," criteria, and weights for the criteria. From these, standards for evaluating the proposals are set up. General assessment areas include:

- Understanding requirements.
- Complying with requirements.
- Ensuring soundness of approach.
- Minimizing risk.
- Evaluating experience and past performance.

Scoring proposals

When you receive the proposals, score them against the standards. Examples of qualitative and quantitative standards are illustrated in TABLES 20-1 and 20-2 respectively. Numerical scoring, such as shown in TABLE 20-3, can be used. Be aware of these pitfalls, however:

- Compare proposals to the standards, not to the other proposals.
- Don't let a deficiency in one area cause you to grade harder in other areas.

- Evaluate one section at a time across all the proposals. As you fatigue and become irritable, you will grade more harshly. The last proposal you look at usually receives harder grading.
- Be aware of your biases. Evaluation is subjective.

Table 20-1. Qualitative Standard Example.

Area: Technical
Item: System integration
Factor: System safety

Description:

The proposed system safety program will be evaluated for adequacy in effecting the design of changes or modifications to the baseline system to achieve special safety objectives. The evaluation will consider the specific tasks, procedures, criteria, and techniques the supplier proposes to use in the system safety program.

Standard:

The standard is met when the proposal:

a. Defines the scope of the system safety effort and supports the stated safety objectives.

b. Defines the qualitative analysis techniques proposed for identifying hazards to the depth required.

c. Describes procedures by which engineering drawings, specifications, test plans, procedures, test data, and results are reviewed at appropriate intervals to ensure safety requirements are specified and followed.

Source: AF Regulation 70-15.

Table 20-2. Quantitative Standard Example.

Area: Operational utility
Item: Mission performance characteristics
Factor: Payload/range

Description:

This factor is defined as the payload that can be carried, considering the basic design gross weight, in a given range, when operational utilization of the aircraft is considered. (Load Factor 2.5)

Standard:

At a weight not exceeding the basic design gross weight, the aircraft is capable of transporting a payload of:

a. 30,000 lbs for a 2800 nm distance.

b. 48,000 lbs for a 1400 nm distance.

Source: AF Regulation 70-15.

Table 20-3. Scoring example.

Numerical Score	Definition
10 9 8 7	Exceptional—exceeds specified performance in a beneficial way, high probability of success, no significant weaknesses.
6 5 4 3	Acceptable—meets standards, good probability of success, weaknesses can be readily corrected.
2 1	Marginal—fails to meet standards, low probability of success, significant deficiencies but correctable.
0	Unacceptable—fails to meet minimum requirements, needs a major change to make it correct.

Source: AF Regulation 70-15.

Assessing the risks

To identify and assess the risks associated with each proposal, you might want to use the following definitions:

High—Likely to cause significant, serious disruption of schedule, increase in cost, or degradation of performance even with special supplier emphasis and close buyer monitoring.

Moderate—Can potentially cause some disruption of schedule, increase in cost, or degradation of performance. Special supplier emphasis and close buyer monitoring will probably be able to overcome difficulties, however.

Low—Has little potential to cause disruption of schedule, increase cost, or degradation of performance. Normal supplier effort and normal buyer monitoring will probably be able to overcome difficulties.

Risks are associated with cost, schedule, and performance. Risk can occur as a result of the technical approach, processes selected, new technology, lack of experience, or other causes.

Evaluating suppliers

To determine whether a supplier can do the job, look for these things in the proposal:

- Does the supplier understand the situation, the problem?
- Has the supplier told you the solution?
- Are the technical, management, and cost sections consistent?

- Are the technical sections consistent with the supplier's risk assessment?
- Does the supplier have prior experience or history of performance?
- Is the supplier committed to quality and delivery?

Summary

Writing proposals

- Your objective for the technical proposal is selling the evaluator. A proposal is a sales tool.
- Address problems in three parts:
    - ~ Explain the problem from the customer's viewpoint.
    - ~ Show understanding of the problem.
    - ~ Give the solution.
- Always discuss the benefits of your product.
- Substantiate your claims.

Evaluating proposals

- The value system has the "must have's," criteria, and weights for the criteria.
- Compare proposals to the standards, not to other proposals..
- Identify and assess the risks associated with each proposal.
- Look for these things:
    - ~ Does the supplier understand the situation, the problem?
    - ~ Has the supplier told you the solution?
    - ~ Is the supplier committed to quality and delivery?

Further reading

Helgeson, Donald V. *Handbook For Writing Technical Proposals That Win Contracts*. Englewood Cliffs, NJ: Prentice-Hall, 1985.

Holtz, Herman. *Government Contracts: Proposalmanship and Winning Strategies*. New York: Plenum Press, 1979.

Kaplan, Marshall H. *Acquiring Major Systems Contracts: Bidding Methods and Winning Strategies*. New York: John Wiley & Sons, 1988.

Appendix A
MIL-STD-499A

FOREWORD

MIL-STD-499A(USAF) has been developed to assist Government and contractor personnel in defining the system engineering effort in support of defense acquisition programs. This Standard applies to internal Department of Defense (DOD) system engineering as well as joint Government-industry applications for Government contracts. The term "contractor", as used throughout this Standard, also means "government agency" when acquisition is being done in-house. The fundamental concept of this Standard is to present a single set of criteria against which all may propose their individual internal procedures as a means of satisfying engineering requirements. Economy is thus achieved by permitting a contractor's internal procedures to be used in support of Air Force programs. In those cases where multi-associate contractors are involved or when more specific direction to a contractor is essential, as determined by the program manager, a set of specific engineering task statements tailored to the specific needs of the program may be specified in the Request for Proposal (RFP).

CONTENTS

1.	SCOPE
1.1	Purpose
1.2	Application
1.3	Implementation
1.4	Tailoring
2.	REFERENCED DOCUMENTS
3.	DEFINITIONS
3.1	Engineering Management
3.2	Technical Program Planning and Control
3.3	System Engineering Process
3.4	Engineering Specialty Integration
3.5	Technical Performance Measurement
4.	GENERAL CRITERIA
5.	DETAILED REQUIREMENTS
5.1	System Engineering Management Plan (SEMP)
5.1.1	Contractual Provisions
5.1.2	Non-Contractual Provisions
5.2	Review of Contractor's Engineering Management
6.	NOTES
6.1	Relationship of Technical Program Planning to Cost and Schedule Planning
6.2	Relationship of Technical Performance Measurement (TPM) to Cost and Schedule Performance Measurement
6.3	Relationship of Integrate Logistics Support (ILS) to System Engineering
6.4	Minimum Documentation
6.5	Data
10.	APPENDIX A
10.1	Technical Program Planning and Control
10.1.1	Development of Contract Work Breakdown Structure (CWBS) and Specification Tree
10.1.2	Program Risk Analysis
10.1.3	System Test Planning
10.1.4	Decision and Control Process
10.1.5	Technical Performance Measurement (TPM)
10.1.5.1	Parameters
10.1.5.2	Planning
10.1.5.3	Implementation of TPM

10.1.5.4	Relating TPM to Cost and Schedule Performance Measurement
10.1.6	Technical Reviews
10.1.6.1	System Requirements Review(s)
10.1.6.2	System Design Review
10.1.6.3	Preliminary Design Review
10.1.6.4	Critical Design Review
10.1.7	Subcontractor/Vendor Reviews
10.1.8	Work Authorization
10.1.9	Documentation Control
10.2	System Engineering Process
10.2.1	Mission Requirements Analysis
10.2.2	Functional Analysis
10.2.3	Allocation
10.2.4	Synthesis
10.2.5	Logistic Engineering
10.2.5.1	Logistic Support Analysis
10.2.5.1.1	Maintenance Engineer Analysis
10.2.5.1.2	Repair Level Analysis
10.2.5.1.3	Logistic Support Modeling
10.2.6	Life Cycle Cost Analysis
10.2.7	Optimization
10.2.7.1	Trade-off Studies
10.2.7.2	System/Cost Effectiveness Analysis
10.2.7.3	Effectiveness Analysis Modeling
10.2.8	Production Engineering Analysis
10.2.9	Generation of Specifications

254 Appendix A

ENGINEERING MANAGEMENT
1. SCOPE

1.1 Purpose. This standard provides the program manger:
   (a) Criteria for evaluating engineering planning and output.
   (b) A means for establishing an engineering effort and a System Engineering Management Plan (SEMP).
   (c) Task statements that may be selectively applied to an acquisition program.

1.2 Application. This standard may be applied at the discretion of the program manager to any system or major equipment program or project. When this standard is applied on a contract, the prime contractor may, at his option, or as specified by the Government, impose tailored requirements of this standard on subcontractors.

1.3 Implementation. This standard may be used in preparing requirements for inclusion in solicitation documents, contract work statement, and System Engineering Management Plans. It is intended that the provisions of this standard be selectively applied in the following combinations:
   (a) Section 5, or
   (b) Section 5 and selected paragraphs from appendix A.

1.4 Tailoring. Selected and tailored task statements of appendix A may be used by:
   (a) Contractors proposing contractual wording in response to an RFP.
   (b) Program managers in preparation of solicitation documents. In each application of appendix A task statements, this standard will be tailored to the specific characteristics of a particular system, program, project, program phase, and/or contractual structure. Tailoring takes the form of deletion, alteration or addition to the task statements. In tailoring the tasks, the depth of detail and level of effort required, and the intermediate and output engineering data expected must be defined. Subsequent tailoring may be done by the contractor and the government during contract negotiations. The agreement reached on the engineering effort and the SEMP shall be reflected in the resultant contract.

2. REFERENCED DOCUMENTS

2.1 The following documents, of the issue in effect on the date of invitation for bids or request for proposal, form a part of this standard to the extent specified herein:

STANDARDS

MILITARY
MIL-STD-480   Configuration control—Engineering Changes, Deviations, and Waivers

MIL-STD-483 (USAF)   Configuration Management Practices for Systems, Equipment, Munitions, and Computer Programs

MIL-STD-881   Work Breakdown Structure for Defense Material Items

MIL-STD-1521(USAF)   Technical Reviews and Audits for Systems, Equipment, and Computer Programs.

SPECIFICATIONS

MILITARY
MIL-S-83490   Specifications, Types and Forms
OTHER PUBLICATIONS
AFLCM/AFSCM 800-4   Optimum Repair-Level Analysis (ORLA)

(Copies of specifications, standards, drawings, and publications required by suppliers in connection with specific procurement functions should be obtained from the procuring activity or as directed by the contracting officer.)

3. DEFINITIONS

The definitions included in applicable documents listed in Section 2 shall apply. Additional definitions established in this document are listed in subsequent paragraphs.

3.1 <u>Engineering Management</u>. The management of the engineering and technical effort required to transform a military requirement into an operational system. It includes the system engineering required to define the system performance parameters and preferred system configuration to satisfy the requirement, the planning and control of technical program tasks, integration of the engineering specialties, and the management of a totally integrated effort of design engineering, specialty engineering, test engineering, logistics engineering, and production engineering to meet cost, technical performance and schedule objectives.

3.2 <u>Technical Program Planning and Control</u>. The management of those design, development, test, and evaluation tasks required to progress from an operational need to the deployment and operation of the system by the user.

3.3 <u>System Engineering Process</u>. A logical sequence of activities and decisions transforming an operational need into a description of system performance parameters and a preferred system configuration.

3.4 <u>Engineering Specialty Integration</u>. The timely and appropriate intermeshing of engineering efforts and disciplines such as reliability, maintainability, logistics engineering, human factors, safety, value engineering, standardization, transportability, etc., to ensure their influence on system design.

3.5 <u>Technical Performance Measurement</u>. The continuing prediction and demonstration of the degree of anticipated or actual achievement of selected technical objectives. It includes an analysis of any differences among the "achievement to date", "current estimate", and the specification requirement. "Achievement to Date" is the value of a technical parameter estimated or measured in a particular test and/or analysis. "Current Estimate" is the value of a technical parameter predicted to be achieved at the end of the contract within existing resources.

4. <u>GENERAL CRITERIA</u>

The contractor's engineering management shall conform to the following general criteria. These criteria are the basis for evaluation of individual program engineering planning and output.

    (a) <u>Technical Objectives</u>. Technical objectives shall be established for each program so that meaningful relationships among need, urgency, risks, and worth can be established.

    (b) <u>Baselines</u>. Functional, allocated, and product baselines shall be developed progressively. Appropriate specifications shall be prepared in accordance with MIL-STD-490.

    (c) <u>Technology</u>. Specification requirements shall be delineated in light of acceptable technological risks defined by risk assessment.

    (d) <u>Realistic System Values</u>. Realistic Reliability, Maintainability, and other such system values shall be established prior to the full-scale development phase.

    (e) <u>Design Simplicity</u>. The concept of design simplicity and standardization shall be evident.

    (f) <u>Design Completeness</u>. The design shall be complete from a total system element viewpoint (hardware, facilities, personnel, computer programs, procedural data).

    (g) <u>Documentation</u>. The concept of minimum documentation shall be evident. Where possible stipulated plans, reports, and other data items shall be used to record the engineering outputs. The repository of this accumulated data will be defined. Engineering data shall be the sole source of performance requirement used in the design and production of the system.

    (h) <u>Engineering Decision Studies</u>. Engineering decisions regarding design alternatives and the technical program shall reflect consideration of system cost effectiveness analysis based on the specified figure(s) of merit, performance parameters, program schedule, resource constraints, producibility, and life cycle cost factors.

    (i) <u>Cost Estimates</u>. Cost estimates shall include acquisition and ownership costs. This shall include any established "design to" cost goals and a current estimate of these costs.

    (j) <u>Technical Task and Work Breakdown Structure Compatibility</u>. Elements of the Contract Work Breakdown Structure and associated technical tasks shall be identified and controlled in accordance with this Standard and MIL-STD-881.

(k) <u>Consistency and Correlation of Requirements</u>. System and technical program requirements shall be consistent, correlatable, and traceable throughout the Contract Work Breakdown Structure so that the impact of technical problems can be promptly determined and accurately appraised.

(l) <u>Technical Performance Measurement</u>. Progress in achieving technical requirements shall be continually assessed. Problems and risk areas shall be identified.

(m) <u>Interface Design Compatibility</u>. Intra-system and intersystem design compatibility of engineering interfaces shall be delineated as interface requirements in appropriate specifications. Interface control requirements and drawings related to (1) the major system elements of a prime contractor's contractual responsibility, (2) other equipment, computer programs, facilities, and procedural data furnished by the Government, and (3) other program participants, shall be coordinated, established and maintained (MIL-STD-483 (USAF)). Clear lines of communication and timely dissemination of changes to these documents shall be maintained.

(n) <u>Engineering Specialty Integration</u>. Engineering efforts such as Integrated Logistic Support (ILS), test engineering, production engineering, transportability, reliability and maintainability engineering, value engineering, safety engineering, electromagnetic compatibility, standardization, etc., shall be integrated into the mainstream design effort.

(o) <u>Engineering Decision Traceability</u>. Significant engineering decisions shall be traceable to the system engineering process activities on which they were based.

(p) <u>Historical Data</u>. Historical engineering/operational data available to system designers shall be identified.

(q) <u>Responsiveness to Change</u>. Changes to system and program requirements in response to directed changes by the procuring activity, or problem solutions identified shall be evaluated for total program impact with respect to performance, cost and schedules.

(r) <u>Compatibility with Related Activities</u>. Engineering Management activities shall be compatible with related program management activities such as cost schedule control system criteria, contract administration, production management, etc.

5. DETAILED REQUIREMENTS. A fully integrated engineering effort meeting the general criteria of Section 4 shall be planned and executed.

5.1 <u>System Engineering Management Plan (SEMP)</u>. A System Engineering Management Plan for satisfying the requirements of this Standard shall be submitted as a separate and complete entity within the contractor's proposal. The plan shall be comprehensive and describe how a fully integrated engineering effort will be managed and conducted. The SEMP shall be in three parts:

Part I

<u>Technical Program Planning and Control</u>. This portion of the plan shall identify organizational responsibilities and authority for system engi-

neering management, including control of subcontracted engineering; levels of control established for performance and design requirements and the control method to be used; technical program assurance methods; plans and schedules for design and technical program reviews; and control of documentation.

Part II
    System Engineering Process. The plan shall contain a detailed description of the process to be used, including the specific tailoring of the process to the requirements of the system and project; the procedures to be used in implementing the process; in-house documentation; the trade study methodology; the types of mathematical and/or simulation models to be used for system and cost effectiveness evaluations; and the generation of specifications.

Part III
    Engineering Specialty Integration. The integration and coordination of the program efforts for the engineering specialty areas, to achieve a best mix of the technical/performance values incorporated in the contract, shall be described in the SEMP with the detailed specialty program plans being summarized or referenced, as appropriate. The SEMP shall depict the integration of the specialty efforts and parameters into the system engineering process and show their consideration during each iteration of the process. Where the specialty programs overlap, the SEMP shall define the responsibilities and authorities of each.

5.1.1 Contractual Provisions. The contractor shall indicate the items in his SEMP which are proposed for inclusion in the contract. Only those items which are basic to the satisfaction of program objectives and the applicable portions of this Standard will be placed on contract.

5.1.2 Non-Contractual Provisions. The contractor shall identify in his SEMP in-house procedures and other planning baselines in sufficient detail to support the procuring activity need for visibility, validation, and verification of the contractual items. Non-contractual items will normally include the details of the engineering organization and key personnel, and other coverage not appropriate for contract change control by the procuring activity.

5.2 Review of Contractor's Engineering Management. Upon request of the procuring activity, the contractor shall make available his engineering management procedures and data for review to determine his capability to satisfy the requirements of this standard and the SEMP. The review shall consist of a combined demonstration and analysis of those features of the contractor's procedures which are key to the satisfaction of the requirements of the contract.

6. NOTES

6.1 Relationship of Technical Program Planning to Cost and Schedule Planning. The technical program planning function defines the detailed planning requirements. It forms the basis for allocation of resources, scheduling of task elements, assignments of authority and responsibility, and the timely integration of all aspects of the technical program. This planning function is carried out to the prescribed contractual levels and integrated with the cost and scheduled control system criteria. The allocated resources becomes the budgeted cost. This relationship pertains both to initial program definition and to the redefinition which is a part of the decisions and control process. (See 10.1.4).

6.2 Relationship of Technical Performance Measurement (TPM) to Cost and Schedule Performance Measurement. The purpose of performance measurement is to: (1) provide visibility of actual vs planned performance, (2) provide early detection or prediction of problems which require management attention, and (3) support assessment of the program impact of proposed change alternatives. TPM assesses the technical characteristics of this system and identifies problems through engineering analyses or tests which indicate performance being achieved for comparison with performance values allocated or specified in contractual documents. Cost/schedule performance measurement assesses the program effort from the point of view of the schedule of increments of work and the cost of accomplishing those increments. By comparing the planned value of work accomplished with both the planned value of work scheduled and the actual cost of work accomplished, problems may surface in the schedule and cost areas. In addition to problems due to unrealistic cost and schedule planning, cost/schedule performance measurement may show up technical inadequacies, just as technical problems identified through TPM can surface inadequacies in budget of time or dollars. Basically, however, cost and schedule performance measurement assumes adequacy of design to meet technical requirements of the system element under consideration; TPM is the complementary function to verify such adequacy. Further, by assessing design adequacy, TPM can deal with the work planned to complete major design and development milestones which need to be changed and thereby provide the basis for forecasting cost and schedule impacts. TPM assessment points should be planned to coincide with the planned completion of significant design and development tasks, or aggregation of tasks. This will facilitate the verification of the results achieved in the completed task in terms of its technical requirements. Thus, TPM and cost/schedule performance measurement are complementary in serving the purpose of program performance measurement.

6.3 Relationship of Integrated Logistic Support (ILS) to System Engineering. ILS planning impacts upon and in turn is impacted by the engineering activities throughout a system life cycle. Initially, support descriptors in the form of criteria and constraints are furnished with the top level system operational needs. These descriptors will include such items as basing concepts, personnel, or training constraints, repair level constraints, and similar support considerations. ILS descriptors should be quantified whenever possible and then be continually and progressively refined and expanded with the evolution of the design. System engineering, in its evolution of functional and detail design requirements, has as its goal the achievement of proper balance among operational, economic, and logistic factors. This balancing and integrating function is an essential part of the system/cost effectiveness trade-offs and studies. Normally, the lower ILS descriptors will influence and be influenced by their relationship to costs of ownership and reliability and Maintainability (R&M) parameters. Thus, the integration of ILS concepts and planning considerations into the system engineering process is a continual and iterative activity, with the output being the optimal balance between performance and support considerations and optimal trade-offs among costs of ownership, schedule, and system effectiveness.

6.4 <u>Minimum Documentation</u>. The iterative nature of the engineering process requires a continual flow of information and documentation. Contractor management information/program control systems, and reports emanating therefrom, shall be utilized to the maximum extent practicable. Proposed changes to existing systems shall consist of only those necessary to satisfy established engineering requirements.

6.5 <u>Data</u>. Selected data items in support of this standard will be reflected in a Contractor Data Requirements List (DD Form 1423), supported by Data Item Descriptions (DD Form 1664) attached to the request for proposal, invitation for bid, or the contract, as appropriate.

Custodians:
   Air Force - 10

Review Activities:
   Air Force - 10, 11, 13, 18, 19

Preparing Activity:
   Air Force - 10
   Project No. MISC-0814

Appendix A
Task Statements

10. This non-mandatory appendix provides specific tasks which may be selected to fit program needs. The scope and depth of the specific tasks chosen for application shall be consistent with the needs of the program. Following their adjustment to specific program needs and subsequent contract negotiations, the following tasks may become specific contractual requirements.

10.1 <u>Technical Program Planning and Control</u>.

10.1.1 <u>Development of Contract Work Breakdown Structure (CWBS) and Specification Tree</u>. The contractor's engineering activity shall develop the technical elements of the Contract Work Breakdown Structure. He shall also prepare a specification tree that relates to his CWBS (MIL-STD-881).

10.1.2 <u>Program Risk Analysis</u>. The program definition and redefinition effort shall include a continuing analysis of the risks associated with the related cost, schedule, and technical parameters. This analysis shall identify critical areas and shall further investigate methods for system or hardware proofing, prototyping, testing, and backup development. The program risk analysis shall also identify test requirements, technical performance measurement parameters, and critical milestones.

10.1.3 <u>System Test Planning</u>. The objectives, scope, and type of system testing shall be products of the engineering effort wherein all engineering specialties are integrated to define an effective and economical total system test program. Whenever practicable, tests for different objectives shall be combined. Test data that is useful for TPM analysis shall be identified and integrated with program planning functions for maximum utility in updating and verifying the technical parameters being tracked. Verification of the acceptability and compatibility of human performance requirements, personnel selection, training, and man-machine interfaces of system procedural data shall also be integrated into the system test program.

10.1.4 <u>Decision and Control Process</u>. Technical, budgetary, and scheduling problems shall be diagnosed as early as possible to determine their impact. Problems and solution alternatives shall be studied to derive the overall impact upon the technical program to ensure that the alternatives are assessed with regard to consideration of side effects that may be induced by the solution. Problem solutions involving changes to the contract requirements or configuration baselines shall be processed in accordance with the change control procedures of the contract.

10.1.5 <u>Technical Performance Measurement (TPM)</u>. A TPM effort tailored to meet specific program needs shall be planned and executed.

10.1.5.1 Parameters. The technical performance parameters selected for tracking shall be key indicators of program success. TPM parameter inter-relationships shall be depicted through construction of tiered dependency trees similar to the specification tree. Each parameter thus identified shall be correlated with a specific CWBS element. Parameters to be reported shall be selected from the total parameters tracked and shall be identified in the SEMP.

10.1.5.2 Planning. The following data, as appropriate, shall be established during the planning stage of this task for each parameter to be tracked:
    (a) Specification requirement.
    (b) Time-phased planned value profile with a tolerance band. The planned value profile shall represent the expected growth of the parameter being tracked. The boundaries of the tolerance band shall represent the inaccuracies of estimation at the time of the estimation, and shall also indicate the region within which it is expected that the specification requirement will be achieved within allocated budget and schedule.
    (c) Program events significantly related to the achievement of the planned value profile.
    (d) Conditions of measurement (type of test, simulation, analysis, etc.).

10.1.5.3 Implementation of TPM. As the design and development activity progresses, the "achievement to date" shall be tracked continually for each of the selected technical performance parameters. In case the "achievement to date" value falls outside the tolerance band, a new profile or "current estimate" shall be developed immediately. The "current estimate" shall be determined from the "achievement to date" and the remaining schedule and budget. The variation shall be determined by comparing the "achievement to date" against the corresponding value on the planned value profile. An analysis shall be accomplished on the variation to determine the causes and to assess the impact on higher level parameters, on interface requirements, and on system cost effectiveness. For technical performance deficiencies, alternate recovery plans shall be developed with cost schedule, and technical performance implications fully explored. For performance in excess of requirements, opportunities for reallocation of requirements and resources shall be assessed.

10.1.5.4 Relating TPM to Cost and Schedule Performance Measurement. The contractor shall indicate how he proposes to relate TPM to cost and schedule performance measurement. Cost, schedule, and technical performance measurement shall be made against common elements of the contract work breakdown structure.

10.1.6 Technical Review. Technical reviews shall be conducted in accordance with MIL-STD-1521 (USAF) to assess the degree of completion of technical efforts related to major milestones before proceeding with further technical effort. The schedule and plan for conduct of technical reviews shall be included in the contractor's System Engineering Management Plan. The reviews shall be a joint effort by contractor and Government representatives. The contractor shall be chairman of the requirements and of the design reviews and shall assure that decisions made as a result of the design review are implemented. Specific reviews shall be identified in the system engineering Management Plan. The following technical reviews are normally required:

10.1.6.1 System Requirements Review(s). These reviews shall be conducted to ascertain progress in defining system technical requirements and implementing other engineering management activity. The number of such reviews will be determined, by the procuring activity.

10.1.6.2 System Design Review. This review shall be conducted to evaluate the optimization, correlation, completeness, and the risks associated with the allocated technical requirements. Also included is a summary review of the system engineering process which produced the allocated technical requirements and of the engineering planning for the next phase of effort. This review will be conducted when the system definition effort has proceeded to the point where system characteristics are defined and the allocated configuration identification has been established. This review will be in sufficient detail to insure a technical understanding among all participants on (1) the updated or completed system or system segment specification, (2) the completed configuration item (CI) development and critical item specifications, and (3) other system definition efforts, productions, and plans.

10.1.6.3 Preliminary Design Review. This review shall be conducted for each CI or aggregate of CI's to (1) evaluate the progress, technical adequacy, and risk resolution (on a technical, cost, and schedule basis) of the selected design approach, (2) determines its compatibility with performance and engineering specialty requirements of the CI development specification, and (3) establish the existence and compatibility of the physical and functional interfaces among the CI and other items of equipment, facilities, computer programs, and personnel.

10.1.6.4 Critical Design Review. This review shall be conducted for each CI when detail design is essentially complete. The purpose of this review will be to (1) determine that the detail design of the CI under review satisfies the performance and engineering specialty requirements of the CI development specifications, (2) establish the detail design compatibility among the CI and other items of equipment, facilities, computer programs and personnel, (3) assess producibility and CI risk areas (on a technical, cost, and schedule basis), and (4) review the preliminary product specifications.

10.1.7 Subcontractor/Vendor Reviews. The contractor shall assure that equipment developed by sub-contractors is reviewed in accordance with the requirements of this standard. These reviews may be accomplished by the contractor or his subcontractors, as desired. The contractor shall assure that actions required as a result of these design reviews are accomplished. Government participation in subcontractor/vendor reviews shall be as specified by the procuring activity.

10.1.8 Work Authorization. Organizational elements responsible for the technical program effort shall be identified and lines of communication defined for control of resources and accomplishment of specific elements of the CWBS. Detailed work authorization (or work orders) shall be compatible with the cost/schedule control system and shall include technical measures of task accomplishment. These technical measures shall be compatible with the contractor Technical Performance Measurement (TPM) process. Work authorization changes may be only those permitted within the general scope of the contract as set forth therein. The contractor shall inform the cognizant Contract Administration Services (CAS) of work authorization changes made.

10.1.9 Documentation Control. Control of in-house drawings, analysis reports, raw test data, work orders, and other technical data shall be traceable, responsive to changes of requirements, and consistent with the configuration management change control requirements of the contract (MIL-STD-480). These data shall be identified for control purposes in a manner similar to engineering drawings.

10.2 System Engineering Process.

10.2.1 Mission Requirements Analysis. Impacts of the stated system operational characteristics, mission objectives, threat, environmental factors, minimum acceptable system functional requirements, technical performance, and system figure(s) of merit as stipulated, proposed, or directed for change shall be analyzed during the conduct of the contract. These impacts shall be examined continually for validity, consistency, desirability, and attainability with respect to current technology, physical resources, human performance capabilities, life cycle costs, or other constraints. The output of this analysis will either verify the existing requirements or develop new requirements which are more appropriate for the mission.

10.2.2 Functional Analysis. System functions and sub-functions shall be progressively identified and analyzed as the basis for identifying alternatives for meeting system performance and design requirements. System functions as used above include the mission, test, production, deployment, and support functions. All contractually specified modes of operational usage and support shall be considered in the analysis. System functions and sub-functions shall be developed in an iterative process based on the results of the mission analysis, the derived system performance requirements, and the synthesis of lower-level system elements. Performance requirements shall be established for each function and sub-function identified. When time is critical to a performance requirement, a time line analysis shall be made.

10.2.3 Allocation. Each function and sub-function shall be allocated a set of performance and design requirements. These requirements shall be derived concurrently with the development of functions, time-line analyses, synthesis of system design, and evaluation performed through trade-off studies and system/cost effectiveness analysis. Time requirements which are prerequisites for a function or set of functions affecting mission success, safety, and availability shall be derived. The derived requirements shall be stated in sufficient detail for allocation to hardware, computer programs, procedural, data, facilities, and personnel. When necessary, special skills or peculiar requirements will be identified. Allocated requirements shall be traceable through the analysis by which they were derived to the system requirement they are designed to fulfill.

10.2.4 <u>Synthesis</u>. Sufficient preliminary design shall be accomplished to confirm and assure completeness of the performance and design requirements allocated for detail design. The performance, configuration, and arrangement of a chosen system and its elements and the technique for their test, support, and operation shall be portrayed in a suitable form such as a set of schematic diagrams, physical and mathematical models, computer simulations, layouts, detailed drawings, and similar engineering graphics. These portrayals shall illustrate intra-and inter-system and item interfaces, permit traceability between the elements at various levels of system detail, and provide means for complete and comprehensive change control. This portrayal shall be the basic source of data for developing, updating, and completing (a) the system, configuration item, and critical item specifications; (b) interface control documentation; (c) consolidated facility requirements; (d) content of procedural handbooks, placards, and similar forms of instructional data; (e) task loading of personnel; (f) operational computer programs; (g) specification trees; and (h) dependent elements of work breakdown structures.

10.2.5 <u>Logistic Engineering</u>. The contractor shall perform logistic engineering as a part of the mainstream engineering effort to develop and achieve a supportable and cost-effective system. This effort will result in establishing the optimal logistic requirements for the deployment and operational phases of the program.

10.2.5.1 <u>Logistic Support Analysis</u>. The contractor shall conduct logistic support analysis leading to the definition of support needs (e.g., maintenance equipment, personnel, spares, repair parts, technical orders, manuals, transportation and handling, etc.). These analyses shall address all levels of operations and maintenance and shall result in requirements for support.

10.2.5.1.1 <u>Maintenance Engineering Analysis</u>. The contractor shall conduct a Maintenance-Engineering Analysis (MEA) which facilitates (a) systematic and complete development of maintenance requirements; (b) sorting and combining logistics data; (c) determination of the quantity of maintenance equipment, personnel, and spares; (d) inputs to system effectiveness and life cycle cost analysis in terms of required factors; and (e) identification of system calibration and measuring standard requirements.

10.2.5.1.2 <u>Repair Level Analysis</u>. The contractor shall conduct a repair level analysis in accordance with AFLCM/AFSCM 800-4. The criteria for conduct of this analysis shall be consistent with the system maintenance concept.

10.2.5.1.3 <u>Logistic Support Modeling</u>. The contractor shall evaluate the impact of support alternatives upon system/equipment life cycle cost, availability, equipment and manpower loading, and stocking of parts shall be predicted and evaluated using modeling techniques when appropriate to the program. The logistic model(s) shall be compatible with and shall not duplicate other system engineering models. Specific models and manual procedures may be identified or provided by the procuring activity.

10.2.6 <u>Life Cycle Cost Analysis</u>. The contractor shall perform and periodically update life cycle cost analyses to include the cost of acquisition and ownership. This effort will result in an identification of the economic consequences of equipment design alternatives.

10.2.7 <u>Optimization</u>. Optimization shall take into consideration the associated risks, technical performance, schedule, and life cycle costs.

10.2.7.1 <u>Trade-off Studies</u>. Desirable and practical trade-offs among stated operational needs, engineering design, program schedule and budget, producibility, supportability, and life cycle costs, as appropriate, shall be continually identified and assessed. Trade-off studies shall be accomplished at the various levels of functional or system detail or as specifically designated to support the decision needs of the system engineering process. Trade-off studies, results and supporting rationale shall be documented in a form consistent with the impact of the study upon program and technical requirements.

10.2.7.2 <u>System/Cost Effectiveness Analysis</u>. A continuing system/cost effectiveness analysis shall be conducted to ensure that engineering decisions, resulting from the review of alternatives, are made only after considering their impact on system effectiveness and cost of acquisition and ownership. The contractor shall identify alternatives which would provide significantly different system effectiveness of costs than those based upon contract requirements.

10.2.7.3 <u>Effectiveness Analysis Modeling</u>. System effectiveness model(s) shall be used when they contribute to the decision process. The model(s) shall allow the input parameters to be varied individually so that their relative effect on total system performance and life cycle cost can be determined. Parameters in the effectiveness model(s) shall correlate to parameters expressed in the performance characteristics allocated to system functions. The model(s) and data file shall be maintained, updated, and modified as required.

10.2.8 <u>Production Engineering Analysis</u>. Production engineering analysis shall be an integral part of the system engineering process. It includes producibility analyses, production engineering inputs to system effectiveness, trade-off studies, and life cycle cost analyses and the consideration of the materials, tools, test equipment, facilities, personnel, and procedures which support manufacturing in RDT&E and production. Critical or special producibility requirements shall be identified as early as possible and shall be an input to the program risk analysis. Where critical or special production engineering requirements provide a constraint on the design, these requirements shall be included in applicable specifications. Long lead time items, material limitations, transition from development to production special processes, and manufacturing constraints shall also be considered and documented during the system engineering process. The contractor shall identify and take necessary steps to reduce high-risk manufacturing areas as early as possible.

10.2.9 <u>Generation of Specifications</u>. The system engineering process shall generate system and item configuration specifications for program peculiar items in accordance with MIL-STD-490 and MIL-STD-83490. The specification effort shall be compatible with the configuration management requirements of the program.

Appendix B
*MIL-STD-1521B excerpts

---

*Excerpts from MIL-STD-1521B, Notice 1, Technical Reviews and Audits for Systems, Equipments, and Computer Software. U.S. Department of Defense, Washington, D.C.

MIL-STD-1521B

SECTION 3

DEFINITIONS

TECHNICAL REVIEWS AND AUDITS

3.1 <u>System Requirements Review (SRR)</u>. The objective of this review is to ascertain the adequacy of the contractor's efforts in defining system requirements. It will be conducted when a significant portion of the system functional requirements has been established.

3.2 <u>System Design Review (SDR)</u>. This review shall be conducted to evaluate the optimization, correlation, completeness, and risks associated with the allocated technical requirements. Also included is a summary review of the system engineering process which produced the allocated technical requirements and of the engineering planning for the next phase of effort. Basic manufacturing considerations will be reviewed and planning for production engineering in subsequent phases will be addressed. This review will be conducted when the system definition effort has proceeded to the point where system characteristics are defined and the configuration items are identified.

3.3 <u>Software Specification Review (SSR)</u>. A review of the finalized Computer Software Configuration Item (CSCI) requirements and operational concept. The SSR is conducted when CSCI requirements have been sufficiently defined to evaluate the contractor's responsiveness to and interpretation of the system, segment, or prime item level requirements. A successful SSR is predicated upon the contracting agency's determination that the Software Requirements Specification, Interface Requirements Specification(s), and Operational Concept Document form a satisfactory basis for proceeding into preliminary software design.

3.4 <u>Preliminary Design Review (PDR)</u>. This review shall be conducted for each configuration item or aggregate of configuration items to (1) evaluate the progress, technical adequacy, and risk resolution (on a technical, cost, and schedule basis) of the selected design approach, (2) determine its compatibility with performance and engineering speciality requirements of the Hardware Configuration Item (HWCI) development specification, (3) evaluate the degree of definition and assess the technical risk associated with the selected manufacturing methods/processes, and (4) establish the existence and compatibility of the physical and functional interfaces among the configuration item and other items of equipment, facilities, computer software, and personnel. For CSCIs, this review will focus on: (1) the evaluation of the progress, consistency, and technical adequacy of the selected top-level design and test approach, (2) compatability between software requirements and preliminary design, and (3) on the preliminary version of the

5

MIL-STD-1521B

operation and support documents.

3.5 <u>Critical</u> <u>Design</u> <u>Review</u> (CDR). This review shall be conducted for each configuration item when detail design is essentially complete. The purpose of this review will be to (1) determine that the detail design of the configuration item under review satisfies the performance and engineering specialty requirements of the HWCI development specifications, (2) establish the detail design compatibility among the configuration item and other items of equipment, facilities, computer software and personnel, (3) assess configuration item risk areas (on a technical, cost, and schedule basis), (4) assess the results of the producibility analyses conducted on system hardware, and (5) review the preliminary hardware product specifications. For CSCIs, this review will focus on the determination of the acceptability of the detailed design, performance, and test characteristics of the design solution, and on the adequacy of the operation and support documents.

3.6 <u>Test</u> <u>Readiness</u> <u>Review</u> (TRR). A review conducted for each CSCI to determine whether the software test procedures are complete and to assure that the contractor is prepared for formal CSCI testing. Software test procedures are evaluated for compliance with software test plans and descriptions, and for adequacy in accomplishing test requirements. At TRR, the contracting agency also reviews the results of informal software testing and any updates to the operation and support documents. A successful TRR is predicated on the contracting agency's determination that the software test procedures and informal test results form a satisfactory basis for proceeding into formal CSCI testing.

3.7 <u>Functional</u> <u>Configuration</u> <u>Audit</u> (FCA). A formal audit to validate that the development of a configuration item has been completed satisfactorily and that the configuration item has achieved the performance and functional characteristics specified in the functional or allocated configuration identification. In addition, the completed operation and support documents shall be reviewed.

3.8 <u>Physical</u> <u>Configuration</u> <u>Audit</u> (PCA). A technical examination of a designated configuration item to verify that the configuration item "As Built" conforms to the technical documentation which defines the configuration item.

3.9 <u>Formal</u> <u>Qualification</u> <u>Review</u> (FQR). The test, inspection, or analytical process by which a group of configuration items comprising the system are verified to have met specific contracting agency contractual performance requirements (specifications or equivalent). This review does not apply to hardware or software requirements verified at FCA for the individual configuration item.

MIL-STD-1521B
APPENDIX A
19 Dec 85

10. **System Requirements Review (SSR)**.

10.1 <u>General</u>. The SRRs are normally conducted during the system Concept Exploration or Demonstration and Validation phase. Such reviews may be conducted at any time but normally will be conducted after the accomplishment of functional analysis and preliminary requirements allocation (to operational/maintenance/training Hardware Configuration Items (HWCIs), Computer Software Configuration Items (CSCIs), facility configuration items, manufacturing considerations, personnel and human factors) to determine initial direction and progress of the contractor's System Engineering Management effort and his convergence upon an optimum and complete configuration.

10.2 <u>Purpose</u>. The total System Engineering Management activity and its output shall be reviewed for responsiveness to the Statement of Work and system/segment requirements. Contracting agency direction to the contractor will be provided, as necessary, for continuing the technical program and system optimization.

10.3 <u>Items to be Reviewed</u>. Representative items to be reviewed include the results of the following, as appropriate:

    a. Mission and Requirements Analysis

    b. Functional Flow Analysis

    c. Preliminary Requirements Allocation

    d. System/Cost Effectiveness Analysis

    e. Trade studies (e.g. addressing system functions in mission and support hardware/firmware/software).

    f. Synthesis

    g. Logistics Support Analysis

    h. Specialty Discipline Studies (i.e., hardware and software reliability analysis, maintainability analysis, armament integration, electromagnetic compatibility, survivability/vulnerability (including nuclear), inspection methods/techniques analysis, energy management, environmental considerations).

    i. System Interface Studies

    j. Generation of Specification

    k. Program Risk Analysis

    l. Integrated Test Planning

Supersedes page 19 of 4 June 1985

MIL-STD-1521B
4 June 1985
APPENDIX A

    m. Producibility Analysis Plans

    n. Technical Performance Measurement Planning

    o. Engineering Integration

    p. Data Management Plans

    q. Configuration Management Plans

    r. System Safety

    s. Human Factors Analysis

    t. Value Engineering Studies

    u. Life Cycle Cost Analysis

    v. Preliminary Manufacturing Plans

    w. Manpower Requirements/Personnel Analysis

    x. Milestone Schedules

10.3.1 The contractor shall describe his progress and problems in:

10.3.1.1 Risk identification and risk ranking (the interrelationship among system effectiveness analysis, technical performance measurement, intended manufacturing methods, and costs shall be discussed, as appropriate).

10.3.1.2 Risk avoidance/reduction and control (the interrelationships with trade-off studies, test planning, hardware proofing, and technical performance measurement shall be discussed, as appropriate).

10.3.1.3 Significant trade-offs among stated system/segment specification requirements/constraints and resulting engineering design requirements/constraints, manufacturing methods/process constraints, and logistic/cost of ownership requirements/constraints and unit production cost/design-to-cost objectives.

10.3.1.4 Identifying computer resources of the system and partitioning the system into HWCIs and CSCIs. Include any trade-off studies conducted to evaluate alternative approaches and methods for meeting operational needs and to determine the effects of constraints on the system. Also include any evaluations of logistics, technology, cost, schedule, resource limitations, intelligence estimates, etc., made to determine their impact on the system. In addition, address the following specific trade-offs related to computer resources:

(Reprinted without change)

MIL-STD-1521B
APPENDIX A

    a. Candidate programming languages and computer architectures evaluated in light of DoD requirements for approved higher order languages and standard instruction set architectures.

    b. Alternative approaches evaluated for implementing security requirements. If an approach has been selected, discuss how it is the most economical balance of elements which meet the total system requirements.

    c. Alternative approaches identified for achieving the operational and support concepts, and, for joint service programs, opportunities for interservice support.

10.3.1.5 Producibility and manufacturing considerations which could impact the program decision such as critical components, materials and processes, tooling and test equipment development, production testing methods, long lead items, and facilities/personnel/skills requirements.

10.3.1.6 Significant hazard consideration should be made here to develop requirements and constraints to eliminate or control these system associated hazards.

10.3.2 Information which the contractor identifies as being useful to his analysis and available through the contracting agency shall be requested prior to this review (e.g., prior studies, operational/support factors, cost factors, safety data, test plan(s), etc.). A separate SRR may be conducted for each of the operational support subsystems depending upon the nature and complexity of the program.

10.4 <u>Post Review Action.</u> After completing the SRR, the contractor shall publish and distribute copies of Review minutes. The contracting agency officially acknowledges completion of the SRR as indicated in paragraph 4.2.4.

21/22

MIL-STD-1521B
APPENDIX B

20. <u>System Design Review (SDR)</u>.

20.1 <u>General.</u> The SDR shall be conducted to evaluate the optimization, traceability, correlation, completeness, and the risk of the allocated requirements, including the corresponding test requirements in fulfilling the system/segment requirements (the functional baseline). The review encompasses the total system requirements, i.e., operations/maintenance/test/training hardware, computer software, facilities, personnel, preliminary logistic support considerations. Also included shall be a summary review of the System Engineering Management Activities (e.g., mission and requirements analysis, functional analysis, requirements allocation, manufacturing methods/process selection, program risk analysis, system/cost effectiveness analysis, logistics support analysis, trade studies, intra- and inter-system interface studies, integrated test planning, specialty discipline studies, and Configuration Management) which produced the above system definition products. A technical understanding shall be reached on the validity and the degree of completeness of the following information:

a. System/Segment Specification

b. The engineering design/cost of the system (see Section 3, Definitions).

c. Preliminary Operational Concept Document

d. Preliminary Software Requirements Specification

e. Preliminary Interface Requirements Specification(s)

f. As appropriate:

    (1) Prime Item Development Specification

    (2) Critical Item Development Specification

20.2 <u>Purpose.</u> An SDR shall be conducted as the final review prior to the submittal of the Demonstration and Validation Phase products or as the initial Full Scale Development Review for systems not requiring a formal Demonstration and Validation Phase but sufficiently complex to warrant the formal assessment of the allocated requirements (and the basis of these requirements) before proceeding with the preliminary design of HWCIs or the detailed requirements analysis for CSCIs. The SDR is primarily concerned with the overall review of the operational/support requirements (i.e., the mission requirements), updated/completed System/Segment Specification requirements, allocated performance requirements, programming and manufacturing methods/processes/planning, and the accomplishment of the System Engineering Management activities to insure that the definition

MIL-STD-1521B
APPENDIX B

effort products are necessary and sufficient. The purposes of the SDR are to:

20.2.1 Insure that the updated/completed System/Segment Specification is adequate and cost effective in satisfying validated mission requirements.

20.2.2 Insure that the allocated requirements represent a complete and optimal synthesis of the system requirements.

20.2.3 Insure that the technical program risks are identified, ranked, avoided, and reduced through:

   a. Adequate trade-offs (particularly for sensitive mission requirements versus engineering realism and manufacturing feasibility to satisfy the anticipated production quantities of corresponding performance requirements);

   b. Subsystem/component hardware proofing;

   c. A responsive test program; and

   d. Implementation of comprehensive engineering disciplines (e.g., worst case analysis, failure mode and effects analysis, maintainability analysis, producibility analysis and standardization.)

20.2.4 Identify how the final combination of operations, manufacturing, maintenance, logistics and test and activation requirements have affected overall program concepts; quantities and types of equipment, unit product cost (see Section 3, Definitions, paragraph 3.11), computer software, personnel, and facilities.

20.2.5 Insure that a technical understanding of requirements has been reached and technical direction is provided to the contractor.

20.3 <u>Items to be Reviewed.</u> The SDR shall include a review of the following items, as appropriate:

20.3.1 System Engineering Management Activities, e.g.:

   a. Mission and Requirements Analysis

   b. Functional Analysis

   c. Requirements Allocation

   d. System/Cost Effectiveness

   e. Synthesis

MIL-STD-1521B
APPENDIX B
19 Dec 85

    f.   Survivability/Vulnerability (including nuclear)

    g.   Reliability/Maintainability/Availability (R/M/A)

    h.   Electromagnetic Compatibility

    i.   Logistic Support Analysis to address, as appropriate, integrated logistics support including maintenance concept, support equipment concept, logistics support concept, maintenance, supply, software support facilities, etc. (MIL-STD-1388-1 and 2)

    j.   System Safety (emphasis shall be placed on system hazard analysis and identification of safety test requirements)

    k.   Securit;y

    l.   Human Factors

    m.   Transportability (including Packaging and Handling)

    n.   System Mass Properties

    o.   Standardization

    p.   Electronic Warfare

    q.   Value Engineering

    r.   System Growth Capability

    s.   Program Risk Analysis

    t.   Technical Performance Measurement Planning

    u.   Producibility Analysis and Manufacturing

    v.   Life Cycle Cost/Design to Cost Goals

    w.   Quality Assurance Program

    x.   Environmental Conditions (Temperature, Vibration, Shock, Humidity, etc).

    y.   Training and Training Support

    z.   Milestone Schedules

    aa.   Software Development Procedures

20.3.2 Results of significant trade studies, for example:

    a.   Sensitivity of selected mission requirements versus

Supersedes page 25 of 4 June 1985

MIL-STD-1521B
APPENDIX B
19 Dec 1985

realistic performance parameters and cost estimates.

    b. Operations design versus maintenance design, including support equipment impacts.

    c. System centralization versus decentralization

    d. Automated versus manual operation

    e. Reliability/Maintainability/Availability

    f. Commercially available items versus new developments

    g. National Stock Number (NSN) items versus new development

    h. Testability trade studies (Allocation of fault detection/isolation capabilities between elements of built-in test, on board/on-site fault detection/isolation subsystem, separate support equipment, and manual procedures)

    i. Size and weight

    j. Desired propagation characteristics versus reduction interference to other systems (optimum selection frequencies)

    k. Performance/logistics trade studies

    l. Life cycle cost reduction for different computer programming languages

    m. Functional allocation between hardware, software, firmware and personnel/procedures

    n. Life Cycle Cost/system performance trade studies to include sensitivity of performance parameters to cost.

    o. Sensitivity of performance parameters versus cost

    p. Cost versus performance

    q. Design versus manufacturing consideration

    r. Make versus buy

    s. Software development schedule

    t. On-equipment versus off-equipment maintenance tasks, including support equipment impacts

    u. Common versus peculiar support equipment

20.3.3 Updated design requirements for operations/maintenance functions and items.

20.3.4 Updated requirements for manufacturing methods and processes.

Supersedes page 26 of 4 June 1985

MIL-STD-1521B
APPENDIX B

20.3.5 Updated operations/maintenance requirements for facilities.

20.3.6 Updated requirements for operations/maintenance personnel and training.

20.3.7 Specific actions to be performed include evaluations of:

   a. System design feasibility and system/cost effectiveness

   b. Capability of the selected configuration to meet requirements of the System/Segment Specification

   c. Allocations of system requirements to subsystems/configuration items

   d. Use of commercially available and standard parts

   e. Allocated inter- and intra- system interface requirements

   f. Size, weight, and configuration of HWCIs to permit economical and effective transportation, packaging, and handling consistent with applicable specifications and standards

   g. Specific design concepts which may require development toward advancing the state-of-the-art

   h. Specific subsystems/components which may require "hardware proofing" and high-risk long-lead time items

   i. The ability of inventory items to meet overall system requirements, and their compatibility with configuration item interfaces

   j. The planned system design in view of providing multi-mode functions, as applicable

   k. Considerations given to:

      (1) Interference caused by the external environment to the system and the system to the external environment.

      (2) Allocated performance characteristics of all system transmitters and receivers to identify potential intra-system electromagnetic (EM) incompatibilities.

      (3) Non-design, spurious and harmonic system performance characteristics and their effect on electromagnetic environments of operational deployments.

   l. Value Engineering studies, preliminary Value Engineering Change Proposals (VECPs) and VECPs (as applicable).

MIL-STD-1521B
APPENDIX B

20.3.8 Review the Preliminary Operational Concept Document, and sections 1.0, 2.0, 3.0, 5.0, 6.0, and 10.0 of the System/Segment Specification, all available HWCI Development Specifications, preliminary Software Requirements, and Interface Requirements Specifications for format, content, technical adequacy, completeness and traceability/correlation to the validated mission/support requirements. All entries marked "not applicable (N/A)" or "to be determined (TBD)" are identified and explained by the contractor.

20.3.9 Review section 4.0 of the System/Segment Specification, all available hardware Development Specifications, and preliminary Software Requirements and Interface Requirements Specifications for format, content, technical adequacy, and completeness. All available test documentation, including HWCI/subsystem and system test plans, shall be reviewed to insure that the proposed test program satisfies the test requirements of section 4.0 of all applicable specifications. All entries labeled "not applicable (N/A)" or "to be determined (TBD)" in section 4.0 of any applicable specification are identified and explained by the contractor.

20.3.10 Review the system, HWCI, and CSCI design for interaction with the natural environment. If any effect or interaction is not completely understood and further study is required, or it is known but not completely compensated for in the design, the proposed method of resolution shall also be reviewed. All proposed environmental tests shall be reviewed for compatibility with the specified natural environmental conditions.

20.3.11 Maintenance functions developed by the contractor to determine that support concepts are valid, technically feasible, and understood. In particular, attention is given to:

 a. R/M/A considerations in the updated System/Segment Specification

 b. Maintenance design characteristics of the system

 c. Corrective and preventive maintenance requirements

 d. Special equipment, tools, or material required

 e. Requirements or planning for automated maintenance analysis

 f. Item Maintenance Analysis compatibility with required maintenance program when weapon is deployed

 g. Specific configuration item support requirements

 h. Forms, procedures, and techniques for maintenance analysis

MIL-STD-1521B
APPENDIX B
19 Dec 85

    i. Maintenance related trade-off studies and findings (includes commercially available equipment, software fault diagnostic techniques)

    j. Logistic cost impacts

    k. Support procedures and tools for computer software which facilitate software modification, improvements, corrections and updates

    l. Hardness critical items/processes

    m. Support equipment concept.

20.3.12 System compliance with nuclear, non-nuclear and laser hardening requirements. High risk areas or design concepts requiring possible advances of the state-of-the-art as a result of survivability criteria shall be identified, and prepared approach(es) to the problem reviewed. Prepared test programs shall be reviewed for sufficiency and compatibility with the specified threat environment and existing simulation test facilities.

20.3.13 The optimization, traceability, completeness, and risks associated with the allocation technical requirements, and the adequacy of allocated system requirements as a basis for proceeding with the development of hardware and software configuration items. Include any available preliminary Software Requirements and Interface Requirements Specifications.

20.3.14 <u>Manufacturing (HWCIs only)</u>.

20.3.14.1 Production feasibility and risk analyses addressed at the SRR shall be updated and expanded. This effort should review the progress made in reducing production risk and evaluate the risk remaining for consideration in the Full Scale Development Phase. Estimates of cost and schedule impacts shall be updated.

20.3.14.2 Review of the Production Capability Assessment shall include:

20.3.14.2.1 A review of production capability shall be accomplished which will constitute an assessment of the facilities, materials, methods, processes, equipment and skills necessary to perform the full scale development and production efforts. Identification of requirements to upgrade or develop manufacturing capabilities shall be made. Requirements for Manufacturing Technology (MANTECH) programs will also be identified as an element of this production assessment.

20.3.14.3 Present the management controls and the design/manufacturing engineering approach to assure that the equipment is producible.

Supersedes page 29 of 4 June 1985

MIL-STD-1521B
APPENDIX B
19 Dec 85

20.3.14.4 Present a review of trade-off studies for design requirements against the requirement for producibility, facilities, tooling, production test equipment, inspection, and capital equipment for intended production rates and volume.

20.3.14.5 The analysis, assessments and trade-off studies should recommend any additional special studies or development efforts as needed.

20.3.15. **Engineering Data**. Evaluate the contractor's drawing system, reviewing the drafting room manual, the preparation and review procedures, change control procedures, flowdown of requirements to subcontractors and vendors, and other aspects fundamental to the acceptability of Level 3 drawings. If available, review completed drawings from other programs or the normal company product line to determine compliance with the company procedures.

20.4 **Post Review Action**. After completing the SDR, the contractor shall publish and distribute copies of Review Minutes. The contracting agency officially acknowledges completion of the SDR as indicated in paragraph 4.2.4.

Supersedes page 30 of 4 June 1985

MIL-STD-1521B
APPENDIX C

30. <u>Software</u> <u>Specification</u> <u>Review</u> (SSR).

30.1 <u>General.</u> The SSR shall be a formal review of a CSCI's requirements as specified in the Software Requirements Specification and the Interface Requirements Specification(s). Normally, it shall be held after System Design Review but prior to the start of CSCI preliminary design. A collective SSR for a group of configuration items, treating each configuration item individually, may be held when such an approach is advantageous to the contracting agency. Its purpose is to establish the allocated baseline for preliminary CSCI design by demonstrating to the contracting agency the adequacy of the Software Requirements Specification (SRS), Interface Requirements Specification(s) (IRS), and Operational Concept Document (OCD).

30.2 <u>Items</u> <u>to</u> <u>be</u> <u>reviewed.</u> The contractor shall present the following items for review by the contracting agency:

a. Functional overview of the CSCI, including inputs, processing, and outputs of each function.

b. Overall CSCI performance requirements, including those for execution time, storage requirements, and similar constraints.

c. Control flow and data flow between each of the software functions that comprise the CSCI.

d. All interface requirements between the CSCI and all other configuration items both internal and external to the system.

e. Qualification requirements that identify applicable levels and methods of testing for the software requirements that comprise the CSCI.

f. Any special delivery requirements for the CSCI.

g. Quality factor requirements; i.e., Correctness, Reliability, Efficiency, Integrity, Usability, Maintainability, Testability, Flexibility, Portability, Reusability, and Interoperability.

h. Mission requirements of the system and its associated operational and support environments.

i. Functions and characteristics of the computer system within the overall system.

j. Milestone schedules.

k. Updates since the last review to all previously delivered

31

MIL-STD-1521B
APPENDIX C

software related CDRL items.

l. Any actions or procedures deviating from approved plans.

30.3 <u>Post Review Action.</u> After completing the SSR, the contractor shall publish and distribute copies of Review Minutes. The contracting agency officially acknowledges completion of the SSR as indicated in paragraph 4.2.4.

30.3.1 The accomplishment of the SSR shall be recorded on the configuration item Development Record by the contractor (see MIL-STD-483, Appendix VII).

MIL-STD-1521B
APPENDIX D

40. <u>Preliminary</u> <u>Design</u> <u>Review (PDR)</u>

40.1 <u>General.</u> The PDR shall be a formal technical review of the basic design approach for a configuration item or for a functionally related group of configuration items. It shall be held after the hardware Development Specification(s), the Software Top Level Design Document (STLDD), the Software Test Plan (STP), the HWCI Test Plan, and preliminary versions of the Computer System Operator's Manual (CSOM), Software User's Manual (SUM), Computer System Diagnostic Manual (CSDM), and Computer Resources Integrated Support Document (CRISD) are available, but prior to the start of detailed design. For each configuration item the actions described below may be accomplished as a single event, or they may be spread over several events, depending on the nature and the extent of the development of the configuration item, and on provisions specified in the contract Statement of Work. A collective PDR for a group of configuration items, treating each configuration item individually, may be held when such an approach is advantageous to the contracting agency; such a collective PDR may also be spread over several events, as for a single configuration item. The overall technical program risks associated with each configuration item shall also be reviewed on a technical, cost, and schedule basis. For software, a technical understanding shall be reached on the validity and the degree of completeness of the STLDD, STP, and the preliminary versions of the CSOM, SUM, CSDM, and CRISD.

40.2 <u>Items</u> <u>to</u> <u>be</u> <u>Reviewed.</u> The contractor shall present the following for review by the contracting agency:

40.2.1 <u>HWCIs:</u>

a. Preliminary design synthesis of the hardware Development Specification for the item being reviewed.

b. Trade-studies and design studies results (see paragraph 20.3.2 of SDR for a representative listing).

c. Functional flow, requirements allocation data, and schematic diagrams.

d. Equipment layout drawings and preliminary drawings, including any proprietary or restricted design/process/components and information.

e. Environment control and thermal design aspects

f. Electromagnetic compatibility of the preliminary design

g. Power distribution and grounding design aspects

h. Preliminary mechanical and packaging design of consoles,

MIL-STD-1521B
APPENDIX D

      racks, drawers, printed circuit boards, connectors, etc.

i. Safety engineering considerations

j. Security engineering considerations

k. Survivability/Vulnerability (including nuclear) considerations

l. Preliminary lists of materials, parts, and processes

m. Pertinent relability/maintainability/availability data

n. Preliminary weight data

o. Development test data

p. Interface requirements contained in configuration item Development Specifications and interface control data (e.g., interface control drawings) derived from these requirements

q. Configuration item development schedule

r. Mock-ups, models, breadboards, or prototype hardware when appropriate

s. Producibility and Manufacturing Considerations (e.g., materials, tooling, test equipment, processes, facilities, skills, and inspection techniques). Identify single source, sole source, diminishing source.

t. Value Engineering Considerations, Preliminary VECPs and VECPs (if applicable).

u. Transportability, packaging, and handling considerations

v. Human Engineering and Biomedical considerations (including life support and Crew Station Requirements).

w. Standardization considerations

x. Description and characteristics of commercially available equipment, including any optional capabilities such as special features, interface units, special instructions, controls, formats, etc., (include limitations of commercially available equipment such as failure to meet human engineering, safety, and maintainability requirements of the specification and identify deficiencies).

y. Existing documentation (technical orders, commercial manuals, etc.,) for commercially available equipment and copies of contractor specifications used to procure

MIL-STD-1521B
APPENDIX D
19 Dec 85

equipment shall be made available for review by the contracting agency.

    z. Firmware to be provided with the system: microprogram logic diagrams and reprogramming/instruction translation algorithm descriptions, fabrication, packaging (integration technology (e.g., LSI, MSI), device types (e.g., CMOS, PMOS)), and special equipment and support software needed for developing, testing, and supporting the firmware.

    aa. Life Cycle Cost Analysis

    ab. Armament compatibility

    ac. Corrosion prevention/control considerations

    ad. Findings/Status of Quality Assurance Program

    ae. Support equipment requirements.

40.2.2 <u>CSCIs:</u>

    a. Functional flow. The computer software functional flow embodying all of the requirements allocated from the Software Requirements Specification and Interface Requirements Specification(s) to the individual Top-Level Computer Software Components (TLCSCs) of the CSCI.

    b. Storage allocation data. This information shall be presented for each CSCI as a whole, describing the manner in which available storage is allocated to individual TLCSCs. Timing, sequencing requirements, and relevant equipment constraints used in determining the allocation are to be included.

    c. Control functions description. A description of the executive control and start/recovery features for the CSCI shall be available, including method of initiating system operation and features enabling recovery from system malfunction.

    d. CSCI structure. The contractor shall describe the top-level structure of the CSCI, the reasons for choosing the components described, the development methodology which will be used within the constraints of the available computer resources, and any support programs which will be required in order to develop/maintain the CSCI structure and allocation of data storage.

    e. Security. An identification of unique security requirements and a description of the techniques to be used for implementing and maintaining security within the CSCI shall be provided.

Supersedes page 35 of 4 June 1985

MIL-STD-1521B
4 June 1985
APPENDIX D

 f. Reentrancy. An identification of any reentrancy requirements and a description of the techniques for implementing reentrant rountines shall be available.

 g. Computer software development facilities. The availability, adequacy, and planned utilization of the computer software development facilities shall be addressed.

 h. Computer software development facility versus the operational system. The contractor shall provide information relative to unique design features which may exist in a TLCSC in order to allow use within the computer software development facility, but which will not exist in the TLCSC installed in the operational system. The contractor shall provide information on the design of support programs not explicitly required for the operational system but which will be generated to assist in the development of the CSCI(s). The contractor shall also provide details of the Software Development Library controls.

 i. Development tools. The contractor shall describe any special simulation, data reduction, or utility tools that are not deliverable under the terms of the contract, but which are planned for use during software development.

 j. Test tools. The contractor shall describe any special test systems, test data, data reduction tools, test computer software, or calibration and diagnostic software that are not deliverable under terms of the contract, but which are planned for use during product development.

 k. Description and characteristics of commercially available computer resources, including any optional capabilities such as special features, interface units, special instructions, controls, formats, etc. Include limitations of commercially available equipment such as failure to meet human engineering, safety and maintainability requirements of the specification and identify deficiencies.

 l. Existing documentation (technical orders, commercial manuals, stc.) for commercially available computer resources and copies of contractor specifications used to procure computer resources shall be made available for review by the contracting agency.

 m. Support resources. The contractor shall describe those resources necessary to support the software and firmware during operational deployment of the system, such as operational and support hardware and software, personnel, special skills, human factors, configuration management, test, and facilities/space.

(Reprinted without change)

288 Appendix B

MIL-STD-1521B
4 June 1985
APPENDIX D

 n. Operation and support documents. The preliminary versions of the CSOM, SUM, CSDM, and CRISD shall be reviewed for technical content and compatability with the top-level design documentation.

 o. Updated since the last review to all previously delivered software related CDRL items.

 p. Review considerations applicable to 40.2.1 as appropriate.

40.2.3 <u>Support Equipment (SE)</u>:

 a. Review considerations applicable to paragraph 40.2.1 and 40.2.2 as appropriate.

 b. Verify testability analysis results. For example, on repairable integrated circuit boards are test points avaliable so that failure can be isolated to the lowest level of repair (See Section 3 Definitions, for "Level of repair").

 c. Verify that the Government furnished SE is planned to be used to the maximum extent possible.

 d. Review progress of long-lead time SE items, identified through interim release and SE Requirements Document (SERD) procedures.

 e. Review progress toward determining total SE requirements for installation, checkout, and test support requirements.

 f. Review the reliability/maintainability/availability of support equipment items.

 g. Identify logistic support requirements for support equipment items and rationale for their selection.

 h. Review calibration requirements.

 i. Describe technical manuals and data availability for support equipment.

 j. Verify compatibility of proposed support equipment with the system maintenance concept.

 k. If a Logistic Support Analysis (LSA) is not done, then review the results of SE trade-off studies for each alternative support concept. For existing SE and printed circuit boards testers, review Maintainability data resulting from the field use of these equipments. Review the cost difference between systems using single or multipurpose SE vs. proposed new SE. Examine technical feasibility in

(Reprinted without change)

MIL-STD-1521B
APPENDIX D
19 Dec 85

using existing, developmental, and proposed new SE. For mobile systems, review the mobility requirements of support equipment.

   l. Review the relationship of the computer resources in the system/subsystem with those in Automatic Test Equipment (ATE). Relate this to the development of Built In Test Equipment (BITE) and try to reduce the need for complex supporting SE.

   m. Verify on-equipment versus off-equipment maintenance task trade study results, including support equipment impacts.

   n. Review updated list of required support equipment.

40.2.4 <u>Engineering Data</u>. Review Level 1 engineering drawings for ease of conversion to higher levels and, if available, review Level 2 and 3 drawings for compliance with requirements. The review of engineering data, as defined in paragraph 3.15, should consider the checklist items discussed in para 100.6, as properly tailored.

40.3 <u>Evaluation of Electrical, Mechanical, and Logical Designs</u>

40.3.1 <u>HWCIs</u>. The material of paragraph 40.2.1 above shall be evaluated to:

   a. Determine that the preliminary detail design provides the capability of satisfying the performance characteristics paragraph of the HWCI Development specifications.

   b. Estalbish compatibility of the HWCI operating characteristics in each mode with overall system design requirements if the HWCI is involved in multi-mode functions.

   c. Establish the existence and nature of physical and functional interfaces between the HWCI and other items of equipment, computer software, and facilities.

40.3.2 <u>CSCIs</u>. The material of paragraph 40.2.2 above shall be evaluated to:

   a. Determine whether all interfaces between the CSCI and all other configuration items both internal and external to the system meet the requirements of the Software Requirements Specification and Interface Requirements Specification(s).

   b. Determine whether the top-level design embodies all the requirements of the Software Requirements Specification and Interface Requirements Specification(s).

Supersedes page 38 of 4 June 1985

MIL-STD-1521B
APPENDIX D
19 Dec 85

    c. Determine whether the approved design methodology has been used for the top-level design.

    d. Determine whether the appropriate Human Factors Engineering (HFE) principals have been incorporated in the design.

    e. Determine whether timing and sizing constraints have been met throughout the top-level design.

    f. Determine whether logic affecting system and nuclear safety has been incorporated in the design.

MIL-STD-1521B
APPENDIX D

40.4 Electromagnetic Compatibility. Review HWCI design for compliance with electromagnetic compatibility/electromagnetic interference (EMC/EMI) requirements. Use Electromagnetic Compatibility Plan as the basis for this review. Check application of MIL-STDs and MIL-Specs cited by the system/equipment specification(s) to the HWCI/Subsystem design. Review preliminary EMI test plans to assess adequacy to confirm that EMC requirements have been met.

40.5 Design Reliability.

40.5.1 Identify the quantitative reliability requirements specified in the hardware Development and Software Requirements Specification(s), including design allocations, and the complexity of the CSCIs.

40.5.2 Review failure rate sources, derating policies, and prediction methods. Review the reliability mathematical models and block diagrams as appropriate.

40.5.3 Describe planned actions when predictions are less than specified requirements.

40.5.4 Identify and review parts or components which have a critical life or require special consideration, and general plan for handling. Agencies so affected shall present planned actions to deal with these components or parts.

40.5.5 Identify applications of redundant HWCI elements. Evaluate the basis for their use and provisions for "on-line" switching of the redundant element.

40.5.6 Review critical signal paths to determine that a fail-safe/fail-soft design has been provided.

40.5.7 Review margins of safety for HWCIs between functional requirements and design provisions for elements, such as: power supplies, transmitter modules, motors, and hydraulic pumps. Similarly, review structural elements; i.e., antenna pedestals, dishes, and radomes to determine that adequate margins of safety shall be provided between operational stresses and design strengths.

40.5.8 Review Reliability Design Guidelines for HWCIs to insure that design reliability concepts shall be available and used by equipment designers. Reliability Design Guidelines shall include, as a minimum, part application guidelines (electrical derating, thermal derating, part parameter tolerances), part selection order of preference, prohibited parts/materials, reliability apportionments/predictions, and management procedures to ensure compliance with the guidelines.

MIL-STD-1521B
APPENDIX D

40.5.9 Review for HWCIs preliminary reliability demonstration plan: failure counting ground rules, accept-reject criteria, number of test articles, test location and environment, planned starting date, and test duration.

40.5.10 Review elements of reliability program plan to determine that each task has been initiated toward achieving specified requirements.

40.5.11 Review subcontractor/supplier reliability controls.

40.6 <u>Design Maintainability</u>

40.6.1 Identify the quantitative maintainability requirements specified in the hardware Development and Software Requirements Specifications; if applicable, compare preliminary predictions with specified requirements.

40.6.2 Review HWCI preventive maintenance schedules in terms of frequencies, durations, and compatibility with system schedules.

40.6.3 Review repair rate sources and prediction methods.

40.6.4 Review planned actions when predictions indicate that specified requirements will not be attained.

40.6.5 Review planned designs for accessibility, testability, and ease of maintenance characteristics (including provisions for automatic or operator-controlled recovery from failure/malfunctions) to determine consistency with specified requirements.

40.6.6 Determine if planned HWCI design indicates that parts, assemblies, and components will be so placed that there is sufficient space to use test probes, soldering irons, and other tools without difficulty and that they are placed so that structural members of units do not prevent access to them or their ease of removal.

40.6.7 Review provisions for diagnosing cause(s) of failure; means for localizing source to lowest replaceable element; adequacy and locations of planned test points; and planned system diagnostics that provide a means for isolating faults to and within the configuration item . This review shall encompass on-line diagnostics, off-line diagnostics, and proposed technical orders and/or commercial manuals.

40.6.8 Review for HWCIs the Design for Maintainability Checklist to insure that listed design principles shall lead to a mature maintainability design. Determine that contractor design engineers are using the checklist.

MIL-STD-1521B
APPENDIX D

40.6.9 Evaluate for HWCIs the preliminary maintainability demonstration plan, including number of maintenance tasks that shall be accomplished; accept-reject criteria; general plans for introducing faults into the HWCI and personnel involved in the demonstration.

40.6.10 Review elements of maintainability program plan to determine that each task has been initiated towards achieving specified requirements.

40.6.11 Insure that consideration has been given to optimizing the system/item from a maintainability and maintenance viewpoint and that it is supportable within the maintenance concept as developed. Also, for HWCIs insure that a Repair Level Analysis (RLA) has been considered.

40.7 Human Factors

40.7.1 The contractor shall present evidence that substantiates the functional allocation decisions. The Review shall cover all operational and maintenance functions of the configuration item. In particular, ensure that the approach to be followed emphasizes the functional integrity of the man with the machine to accomplish a system operation.

40.7.2 Review design data, design descriptions and drawings on system operations, equipments, and facilities to insure that human performance requirements of the hardware Development and Software Requirements Specifications are met. Examples of the types of design information to be reviewed are:

   a. Operating modes for each display station, and for each mode, the functions performed, the displays and control used, etc.

   b. The exact format and content of each display, including data locations, spaces, abbreviations, the number of digits, all special symbols (Pictographic), alert mechanisms (e.g., flashing rates), etc.

   c. The control and data entry devices and formats including keyboards, special function keys, cursor control, etc.

   d. The format of all operator inputs, together with provisions for error detection and correction.

   e. All status, error, and data printouts - including formats, headings, data units, abbreviations, spacings, columns, etc.

These should be presented in sufficient detail to allow contracting agency personnel to judge adequacy from a human usability standpoint, and design personnel to know what is required, and test personnel to prepare tests.

MIL-STD-1521B
APPENDIX D

40.7.3 Make recommendations to update the System/Segment, or Software Requirements Specification and Interface Requirements Specification(s) in cases where requirements for human performance need to be more detailed.

40.7.4 Review man/machine functions to insure that man's capabilities are utilized and that his limitations are not exceeded.

40.8 <u>System Safety</u>

40.8.1 Review results of configuration item safety analyses, and quantitative hazard analyses (if applicable).

40.8.2 Review results of system and intra-system safety interfaces and trade-off studies affecting the configuration item.

40.8.3 Review safety requirements levied on subcontractors.

40.8.4 Review known special areas of safety, peculiar to the nature of the system (e.g., fuel handling, fire protection, high levels of radiated energy, high voltage protection, safety interlocks, etc.).

40.8.5 Review results of preliminary safety tests (if appropriate).

40.8.6 Generally review adequacy and completeness of configuration item from design safety viewpoint.

40.8.7 Review compliance of commercially available configuration items or configuration item components with system safety requirements and identify modifications to such equipment, if required.

40.9 <u>Natural Environment</u>

40.9.1 Review contractor's planned design approach toward meeting climatic conditions (operating and non-operating ranges for temperature, humidity, etc.) that are specified in the HWCI Development Specification.

40.9.2 Insure that the contractor clearly understands the effect of, and the interactions between, the natural aerospace environment and HWCI design. In cases where the effect and interactions are not known or are ambiguous, insure that studies are in progress or planned to make these determinations.

40.9.3 Current and forecast natural aerospace environment parameters may be needed for certain configuration items; e.g., display of airbase conditions in a command and control system, calculation of impact point for a missile, etc. Insure

MIL-STD-1521B
APPENDIX D

compatibility between the configuration item design and appropriate meteorological communications by comparing characteristics of the source (teletype, facsimile, or data link) with that of the configuration item. Insure that arrangements or plans to obtain needed information have been made and that adequate display of natural environmental information shall be provided.

40.10 <u>Equipment and Part Standardization</u>

40.10.1 <u>Equipment and Components</u>:

   a. Review current and planned contractor actions to determine that equipment or components for which standards or specifications exist shall be used whenever practical. (Standard item with NSN should have first preference).

   b. Review specific trade-offs or modifications that may be required of existing designs if existing items are, or will be, incorporated in the HWCI.

   c. Existing designs will be reviewed for use or non-use based on the potential impact on the overall program in the following areas:

   (1) Performance

   (2) Cost

   (3) Time

   (4) Weight

   (5) Size

   (6) Reliability

   (7) Maintainability

   (8) Supportability

   (9) Producibility

   d. Review HWCI design to identify areas where a practical design change would materially increase the number of standard items that could be incorporated.

   e. Insure that Critical Item Specifications shall be prepared for hardware items identified as engineering or logistics critical.

MIL-STD-1521B
APPENDIX D

40.10.2 Parts Standardization and Interchangeability:

a. Review procedures to determine if maximum practical use will be made of parts built to approved standards or specifications. The potential impact on the overall program is to be evaluated when a part built to approved standards and specifications cannot be used for any of the following reasons:

   (1) Performance

   (2) Weight

   (3) Size

   (4) Reliability/Mantainability/Availability

   (5) Supportability

   (6) Survivability (including nuclear)

b. Identify potential design changes that will permit a greater use of standard or preferred parts and evaluate the trade-offs.

c. Insure understanding of parts control program operations for selection and approval of parts in new design or major modifications.

d. Review status of the Program Parts Selection List.

e. Review status of all non-standard parts identified.

f. Review pending parts control actions that may cause program slippages, such as non-availability of tested parts.

40.10.3 Assignment of Official Nomenclature:

a. Insure understanding of procedure for obtaining assignment of nomenclature and approval of nameplates.

b. Determine that a nomenclature conference has been held and agreement has been reached with the contracting agency on the level of nomenclature; i.e., system, set, central, group, component, sub-assembly, unit, etc.

40.11 Value Engineering

40.11.1 Review the Contractor's in-house incentive Value Engineering Program, which may include but not be limited to the following:

MIL-STD-1521B
APPENDIX D

a. Contractor's Value Engineering organization, policies and procedures.

b. Contractor's Value Engineering Training Program.

c. Potential Value Engineering projects, studies and VECPs.

d. Schedule of planned Value Engineering tasks/events.

e. Policies and procedures for subcontractor Value Engineering Programs.

40.12 Transportability

40.12.1 Review HWCI to determine if design meets contracts requirements governing size and weight to permit economical handling, loading, securing, transporting, and disassembly for shipment within existing capabilities of military and commercial carriers. Identify potential outsized and overweight items. Identify system/items defined as being hazardous. Ensure packaging afforded hazardous items complies with hazardous materials regulations.

40.12.2 Identify HWCIs requiring special temperature and humidity control or those possessing sensitive and shock susceptibility characteristics. Determine special transportation requirements and availability for use with these HWCIs.

40.12.3 Review Transportability Analysis to determine that transportation conditions have been evaluated and that these conditions are reflected in the design of protective, shipping, and handling devices. In addition to size and weight characteristics, determine that analysis includes provisions for temperature and humidity controls, minimization of sensitivity, susceptibility to shock, and transit damage.

40.13 Test

40.13.1 Review all changes to the System/Segment, HWCI Development, Software Requirements, and Interface Requirements Specifications subsequent to the established Allocated Baseline to determine whether Section 4.0 of all these specifications adequately reflects these changes.

40.13.2 Review information to be provided by the contractor regarding test concepts for Development Test and Evaluation (DT&E) testing (both informal and formal). Information shall include:

a. The organization and responsibilities of the group that will be responsible for test.

MIL-STD-1521B
APPENDIX D

b. The management of his in-house development test effort provides for:

   (1) Test Methods (plans/procedures)

   (2) Test Reports

   (3) Resolution of problems and errors

   (4) Retest procedure

   (5) Change control and configuration management

   (6) Identification of any special test tools that are not deliverable under the contract.

c. The methodology to be used to meet quality assurance requirements/qualification requirements, including the test repeatability characteristics and approach to regression testing.

d. The progress/status of the test effort since the previous reporting milestone.

40.13.3 Review status of all negative or provisional entries such as "not applicable (N/A)" or "to be determined (TBD)" in Section 4.0 of the System/Segment, hardware Development, Software Requirements or Interface Requirements Specifications. Review all positive entries for technical adequacy. Insure that associated test documentation includes these changes.

40.13.4 Review interface test requirements specified in Section 4.0 of the hardware Development, Software Requirements, and Interface Requirements Specifications for compatibility, currency, technical adequacy, elimination of redundant test. Insure that all associated test documents reflect these interface requirements.

40.13.5 Insure that all test planning documentation has been updated to include new test support requirements and provisions for long-lead time support requirements.

40.13.6 Review contractor test data from prior testing to determine if such data negates the need for additional testing.

40.13.7 Examine all available breadboards, mock-ups, or devices which will be used in implementing the test program or which affect the test program, for program impact.

40.13.8 Review plans for software Unit testing to ensure that they:

MIL-STD-1521B
APPENDIX D

a. Address Unit level sizing, timing, and accuracy requirements.

b. Present general and specific requirements that will be demonstrated by Unit testing.

c. Describe the required test-unique support software, hardware, and facilities and the interrelationship of these items.

d. Describe how, when, and from where the test-unique support items will be obtained.

e. Provide test schedules consistent with higher level plans.

40.13.9 Review plans for CSC integration testing to ensure that they:

a. Define the type of testing required for each level of the software structure above the unit level.

b. Present general and specific requirements that will be demonstrated by CSC integration testing.

c. Describe the required test-unique support software, hardware, and facilities and the interrelationship of these items.

d. Describe how, when, and from where the test-unique support items will be obtained.

e. Describe CSC integration test management, to include:

   (1) Organization and responsibilities of the test team

   (2) Control procedures to be applied during test

   (3) Test reporting

   (4) Review of CSC integration test results

   (5) Generation of data to be used in CSC integration testing.

f. Provide test schedules consistent with higher level plans.

40.13.10 Review plans for formal CSCI testing to ensure that they:

a. Define the objective of each CSCI test, and relate the test to the software requirements being tested.

MIL-STD-1521B
APPENDIX D

b. Relate formal CSCI tests to other test phases.

c. Describe support software, hardware, and facilities required for CSCI testing; and how, when, and from where they will be obtained.

d. Describe CSCI test roles and responsibilities.

e. Describe requirements for Government-provided software, hardware, facilities, data, and documentation.

f. Provide CSCI test schedules consistent with higher-level plans.

g. Identify software requirements that will be verified by each formal CSCI test.

40.14 <u>Maintenance and Maintenance Data (HWCIs)</u>

40.14.1 Describe System Maintenance concept for impact on design and SE. Review adequacy of maintenance plans. Coverage shall be provided for On Equipment (Organizational), Off Equipment - On Site (Intermediate), Off Equipment - Off Site (Depot) level maintenance of Government Furnished Equipment (GFE), and Contractor Furnished Equipment (CFE). (See Section 3, Definitions, para 3.12 for levels of maintenance.)

40.14.2 Determine degree of understanding of the background, purpose, requirements, and usage of Maintenance (failure) Data Collection and Historical/Status Records. (Ref Data Item titled, "Reliability and Maintainability Data Reporting and Feedback Failure Summary Reports").

40.14.3 Describe method of providing Maintenance, Failure, Reliability, Maintainability Data to contracting agency.

40.14.4 Describe how requirements are submitted to the contracting agency for Equipment Classification (EQ/CL) Codes (formerly Work Order Number Prefix/Suffix Codes) when this requirement exists.

40.14.5 Review plans for (and status of) Work Unit Coding of the equipment. Work Unit codes shall be available for documenting Maintenance Data commencing with configuration item/Subsystem Testing. (Ref. Data Item titled "Technical Orders" and the military specification on work unit coding).

40.15 <u>Spares and Government Furnished Property (GFP)</u>.

40.15.1 Review logistics and provisioning planning to insure full understanding of scope of requirements in these areas and that a reasonable time-phased plan has been developed for accomplishment. Of specific concern are the areas of: provisioning requirements,

MIL-STD-1521B
APPENDIX D

GFP usage, and spare parts, and support during installation, checkout, and test.

40.15.2 Review provisioning actions and identify existing or potential provisioning problems - logistic critical and long-lead time items are identified and evaluated against use of the interim release requirements.

40.15.3 Review plans for maximum screening and usage of GFP, and extent plans have been implemented.

40.15.4 Review progress toward determining and acquiring total installation, checkout, and test support requirements.

40.16 <u>Packaging/SDPE (Special Design Protective Equipment)</u>

40.16.1 Analyze all available specifications (System/Segment, HWCI Development, Software Requirements, Interface Requirements, and Critical Items) for packaging (Section 5) requirements for each product fabrication and material specification.

40.16.2 Evaluate user/operational support requirements and maintenance concepts for effect and influence on package design.

40.16.3 Establish that time phased plan for package design development is in consonance with the development of the equipment design.

40.16.4 Review planned and/or preliminary equipment designs for ease of packaging and simplicity of package design, and identify areas where a practical design change would materially decrease cost, weight, or volume of packaging required.

40.16.5 Review requirements for SDPE necessary to effectively support configuration item during transportation, handling and storage processes. Insure SDPE is categorized as a configuration item utilizing specifications conforming to the types and forms as prescribed in the contract. Review SDPE development/product specifications for adequacy of performance/interface requirements.

40.16.6 Determine initial package design baselines, concepts, parameters, constraints, etc., to the extent possible at this phase of the configuration item development process.

40.16.7 Insure previously developed and approved package design data for like or similar configuration items is being utilized.

40.16.8 Establish plans for trade studies to determine the most economical and desirable packaging design approach needed to satisfy the functional performance and logistic requirements.

MIL-STD-1521B
APPENDIX D

40.16.9 Verify the adequacy of the prototype package design.

40.16.10 Review Section 5 of Specification to insure full understanding by contractor for contractor requirements. Identify package specification used for hazardous materials.

40.17 Technical Manuals

40.17.1 Review status of the "Technical Manual Publications Plan" to insure that all aspects of the plan have been considered to the extent that all concerned agencies are apprised of the technical manual coverage to be obtained under this procurement. The suitability of available commercial manuals and/or modifications thereto shall also be determined.

40.17.2 Review the availability of technical manuals for validation/verification during the latter phases of DT&E testing.

40.17.3 If a Guidance Conference was not accomplished or if open items resulted from it, then review as applicable provisions for accomplishing TO in-process reviews, validation, verification, prepublication, and postpublication reviews.

40.18 System Allocation Document

40.18.1 Review the Draft System Allocation Document for completeness and technical adequacy to extent completed.

40.18.2 The format shall provide the following minimum information:

   a. Drawing Number
   b. Issue
   c. Number of Sheets
   d. Location
   e. Configuration Item Number
   f. Title
   g. Part Number
   h. Serial Number
   i. Specification Number
   j. Equipment Nomenclature
   k. Configuration Item Quantity
   l. Assembly Drawing

40.19 Design Producibility and Manufacturing

40.19.1 The contractor shall demonstrate and present evidence that manufacturing engineering will be integrated into the design process.

MIL-STD-1521B
APPENDIX D

a. The contractor shall provide evidence of performing producibility analyses on development hardware trading off design requirements against manufacturing risk, cost, production, volume, and existing capability/availability.

Evidence of such analyses may be in the contractor's own format but must conclusively demonstrate that in-depth analyses were performed by qualified organizations/individuals and the results of those analyses will be incorporated in the design.

b. Preliminary manufacturing engineering and production planning demonstrations shall address: material and component selection, preliminary production sequencing, methods and flow concepts, new processes, manufacturing risk, equipment and facility utilization for intended rates and volume, production in-process and acceptance test and inspection concepts. (Efforts to maximize productivity in the above areas should be demonstrated.)

c. Management systems to be utilized will insure that producibility and manufacturing considerations are integrated throughout the FSD effort.

40.19.2 The producibility and manufacturing concerns identified in the SRR and the SDR shall be updated and expanded to:

a. Provide evidence that concerns identified in the Manufacturing Feasibility Assessment and the Production Capability Estimate have been addressed and that resolutions are planned or have been performed.

b. Make recommendations including manufacturing technology efforts and provide a schedule of necessary actions to the program office to resolve open manufacturing concerns and reduce manufacturing risk.

40.20 <u>Post Review Action</u>

40.20.1 After completing the PDR, the contractor shall publish and distribute copies of Review minutes. The contracting agency officially acknowledges completion of a PDR as indicated in paragraph 4.2.4.

40.20.2 The accomplishment of the PDR shall be recorded on the configuration item Development Record by the contractor.

MIL-STD-1521B
APPENDIX E

50. Critical Design Review

50.1 General. The CDR shall be conducted on each configuration item prior to fabrication/production/coding release to insure that the detail design solutions, as reflected in the Draft Hardware Product Specification, Software Detailed Design Document (SDDD), Data Base Design Document(s) (DBDD(s)), Interface Design Document(s) (IDD(s)), and engineering drawings satisfy requirements established by the hardware Development Specification and Software Top Level Design Document (STLDD). CDR shall be held after the Computer Software Operator's Manual (CSOM), Software User's Manual (SUM), Computer System Diagnostic Manual (CSDM), Software Programmer's Manual (SPM), and Firmware Support Manual (FSM) have been updated or newly released. For complex/large configuration items the CDR may be conducted on an incremental basis, i.e., progressive reviews are conducted versus a single CDR. The overall technical program risks associated with each configuration item shall also be reviewed on a technical (design and manufacturing), cost and schedule basis. For software, a technical understanding shall be reached on the validity and the degree of completeness of the SDDD, IDD(s), DBDD(s), STD, CRISD, SPM, and FSM, and preliminary versions of the CSOM, SUM, and CSDM.

50.1.1 Equipment/Facilities configuration items. The detail design as disclosed by the hardware Product Specification, drawings, schematics, mockups, etc., shall be reviewed against the HWCI Development Specification performance requirements. For other than facilities, the result of a successful CDR shall be the establishment of the design baseline for detailed fabrication/production planning i.e., the contractor is permitted to use the detail design as presented at CDR and reflected in the hardware Product Specification for planning for production and, if specifically authorized, for initial fabrication/production efforts.

50.1.2 Computer Software configuration items (CSCIs). The CDR for a CSCI shall be a formal technical review of the CSCI detail design, including data base and interfaces. The CDR is normally accomplished for the purpose of establishing integrity of computer software design at the level of a Unit's logical design prior to coding and testing. CDR may be accomplished at a single review meeting or in increments during the development process corresponding to periods at which components or groups of components reach the completion of logical design. The primary product of the CDR is a formal identification of specific software documentation which will be released for coding and testing. By mutual agreement between the contractor and the contracting agency, CDRs may be scheduled concurrently for two or more CSCIs.

50.1.2.1 Since computer software development is an iterative process, the completion of a CDR for a CSCI is not necessarily sufficient for maintaining adequate visibility into the remaining

MIL-STD-1521B
APPENDIX E

development effort through testing.

50.1.2.2 Additional In-Progress Reviews may be scheduled post-CDR which address:

- a. Response to outstanding action items
- b. Modifications to design necessitated by approved ECPs or design/program errors
- c. Updating sizing and timing data
- d. Updated design information, as applicable
- e. Results obtained during in-house testing, including problems encountered and solutions implemented or proposed.

50.2 <u>Items to be Reviewed.</u> The contractor shall present the following for review by the contracting agency:

50.2.1 <u>HWCIs</u>

- a. Adequacy of the detail design reflected in the draft hardware Product Specification in satisfying the requirements of the HWCI Development Specification for the item being reviewed.
- b. Detail engineering drawings for the HWCI including schematic diagrams.
- c. Adequacy of the detailed design in the following areas:

    (1) Electrical design

    (2) Mechanical design

    (3) Environmental control and thermal aspects

    (4) Electromagnetic compatibility

    (5) Power generation and grounding

    (6) Electrical and mechanical interface compatibility

    (7) Mass properties

    (8) Reliability/Maintainability/Availability

    (9) System Safety Engineering

    (10) Security Engineering

MIL-STD-1521B
APPENDIX E
19 Dec 85

    (11) Survivability/Vulnerability (including nuclear)

    (12) Producibility and Manufacturing

    (13) Transportability, Packaging and handling

    (14) Human Engineering and Biomedical Requirements (including Life Support and Crew Station Requirements)

    (15) Standardization

    (16) Design versus Logistics Trade-offs

    (17) Support equipment requirements

  d. Interface control drawings

  e. Mock-ups, breadboards, and/or prototype hardware

  f. Design analysis and test data

  g. System Allocation Document for HWCI inclusion at each scheduled location.

  h. Initial Manufacturing Readiness (for example, manufacturing engineering, tooling demonstrations, development and proofing of new materials, processes, methods, tooling, test equipment, procedures, reduction of manufacturing risks to acceptable levels).

  i. Preliminary VECPs and/or formal VECPs

  j. Life cycle costs

  k. Detail design information on all firmware to be provided with the system.

  l. Verify corrosion prevention/control considerations to insure materials have been chosen that will be compatible with operating environment.

  m. Findings/Status of Quality Assurance Program

50.2.2 <u>CSCIs</u>.

  a. Software Detailed Design, Data Base Design, and Interface Design Document(s). In cases where the CDR is conducted in increments, complete documents to support that increment shall be available.

  b. Supporting documentation describing results of analyses, testing, etc., as mutually agreed by the contracting agency and the contractor.

Supersedes page 55 of 4 June 1985

MIL-STD-1521B
APPENDIX E
19 Dec 85

c. System Allocation Document for CSCI inclusion at each scheduled location.

d. Computer Resources Integrated Support Document.

e. Software Programmer's Manual

f. Firmware Support Manual

g. Progress on activities required by CSCI PDR (para 40.2.2).

h. Updated operation and support documents (CSOM, SUM, CSDM).

i. Schedules for remaining milestones.

j. Updates since the last review to all previously delivered software related CDRL items.

50.2.3 <u>Support Equipment (SE)</u>:

a. Review requirements (paragraphs 50.2.1 and 50.2.2) for SE.

b. Verify maximum considerations GFE SE

c. Identify existing or potential SE provisioning problems

d. Determine qualitative and quantitative adequacy of provisioning drawings and data

e. Review reliability of SE

f. Review logistic support requirements for SE items

g. Review Calibration requirements

h. Review documentation for SE.

50.2.4. <u>Engineering Data</u>. Continuing from the results of the Preliminary Design Review (PDR), review engineering data as defined in para 3.15, as to suitability for intended use. The review should consider the checklist items discussed in para 100.6, as properly tailored.

50.3 <u>Detailed Evaluation of Electrical, Mechanical, and Logical Designs</u>

50.3.1 <u>HWCIs</u>. Detailed block diagrams, schematics, and logic diagrams shall be compared with interface control drawings to determine system compatibility. Analytical and available test data shall be reviewed to insure the hardware Development Specification has been satisfied.

Supersedes page 56 of 4 June 1985

```
              MIL-STD-1521B
              APPENDIX E
              19 Dec 1985
```

50.3.1.1 The contractor shall provide information on firmware which is included in commercially available equipment or to be included in equipment developed under the contract. Firmware in this context includes the microprocessor and associated sequence of micro-instructions necessary to perform the allocated tasks. As a minimum, the information presented during CDR shall provide

MIL-STD-1521B
APPENDIX E

descriptions and status for the following:

    a. Detailed logic flow diagrams

    b. Processing algorithms

    c. Circuit diagrams

    d. Clock and timing data (e.g., timing charts for micro-instructions)

    e. Memory (e.g., type (RAM, PROM), word length, size (total and spare capacity))

    f. Micro-instruction list and format

    g. Device functional instruction set obtained by implementation of firmware.

    h. Input/output data width (i.e., number of bits for data and control.)

    i. Self-test (diagnostics) within firmware.

    j. Support software for firmware development:

        (1) Resident assembler

        (2) Loader

        (3) Debugging routines

        (4) Executive (monitor)

        (5) Non-resident diagnostics

        (6) Cross assembler and higher level language on host computer

        (7) Instruction simulator

50.3.2 **CSCIs.** The contractor shall present the detailed design (including rationale) of the CSCI to include:

    a. The assignment of CSCI requirements to specific Lower-Level Computer Software Components (LLCSCs) and Units, the criteria and design rules used to accomplish this assignment, and the traceability of Unit and LLCSC designs to satisfy CSCI requirements, with emphasis on the necessity and sufficiency of the Units for implementing TLCSC design requirements.

MIL-STD-1521B
APPENDIX E

b. The overall information flow between software Units, the method(s) by which each Unit gains control, and the sequencing of Units relative to each other.

c. The design details of the CSCI, TLCSCs, LLCSCs, and Units including data definitions, timing and sizing, data and storage requirements and allocations.

d. The detailed design characteristics of all interfaces, including their data source, destination, interface name and interrelationships; and, if applicable, the design for direct memory access. The contractor shall also give an overview of the key design issues of the interface software design, and indicate whether data flow formats are fixed or subject to extensive dynamic changes.

e. The detailed characteristics of the data base. Data base structure and detailed design, including all files, records, fields, and items. Access rules, how file sharing will be controlled, procedures for data base recovery/regeneration from a system failure, rules for data base manipulation, rules for maintaining file integrity, rules for usage reporting, and rules governing the types and depth of access shall be defined. Data management rules and algorithms for implementing them shall be described. Details of the language required by the user to access the data base shall also be described.

50.4 Electromagnetic Compatibility:

a. Review contractor EMC design of all HWCIs. Determine compliance with requirements of the Electromagnetic Compatibility Plan and HWCI specifications.

b. Review system EMC including effects on the electromagnetic environment (inter-system EMC) and intra-system EMC. Determine acceptability of EMC design and progress toward meeting contractual EMC requirements.

c. Review EMC test plans. Determine adequacy to confirm EMC design characteristics of the system/HWCI/subsystem.

50.5 Design Reliability:

50.5.1 Review the most recent predictions of hardware and software reliability and compare against requirements specified in hardware Development Specification and Software Requirements Specification. For hardware, predictions are substantiated by review of parts application stress data.

50.5.2 Review applications of parts or configuration items with minimum life, or those which require special consideration to

MIL-STD-1521B
APPENDIX E

insure their effect on system performance is minimized.

50.5.3 Review completed Reliability Design Review Checklist to insure principles have been satisfactorily reflected in the configuration item design.

50.5.4 Review applications of redundant configuration item elements or components to establish that expectations have materialized since the PDR.

50.5.5 Review detailed HWCI reliability demonstration plan for compatibility with specified test requirements. The number of test articles, schedules, locations, test conditions, and personnel involved are reviewed to insure a mutual understanding of the plan and to provide overall planning information to activities concerned.

50.5.6 Review the failure data reporting procedures and methods for determination of failure trends.

50.5.7 Review the thermal analysis of components, printed circuit cards, modules, etc. Determine if these data are used in performing the detailed reliability stress predictions.

50.5.8 Review on-line diagnostic programs, off-line diagnostic programs, support equipment, and preliminary technical orders (and/or commercial manuals) for compliance with the system maintenance concept and specification requirements.

50.5.9 Review software reliability prediction model and its updates based upon test data and refined predictions of component usage rates and complexity factors.

50.6 Design Maintainability

50.6.1 Review the most recent predictions of quantitative maintainability and compare these against requirements specified in the HWCI Development Specification and Software Requirements Specification.

50.6.2 Review preventive maintenance frequencies and durations for compatibility with overall system requirements and planning criteria.

50.6.3 Identify unique maintenance procedures required for the configuration item during operational use and evaluate their total effects on system maintenance concepts. Assure that system is optimized from a maintenance and maintainability viewpoint and conforms with the planned maintenance concept. This shall include a review of provisions for automatic, semi-automatic, and manual recovery from hardware/software failures and malfunctions.

MIL-STD-1521B
APPENDIX E

50.6.4 Identify design-for-maintainability criteria provided by the checklist in the design detail to insure that criteria have, in fact been incorporated.

50.6.5 Determine if parts, assemblies, and other items are so placed that there is sufficient space to use test probes, soldering irons, and other tools without difficulty and that they are placed so that structural members of units do not prevent access to them or their ease of removal.

50.6.6 Review detailed maintainability demonstration plan for compatibility with specified test requirements. Supplemental information is provided and reviewed to insure a mutual understanding of the plan and to provide overall planning information to activities concerned.

50.7 <u>Human Factors</u>

50.7.1 Review detail design presented on drawings, schematics, mockups, or actual hardware to determine that it meets human performance requirements of the HWCI Development Specification and Software Requirements Specification, Interface Requirements Specification(s), and accepted human engineering practices.

50.7.2 Demonstrate by checklist or other formal means the adequacy of design for human performance.

50.7.3 Review each facet of design for man/machine compatibility. Review time/cost/effectiveness considerations and forced trade-offs of human engineering design.

50.7.4 Evaluate the following human engineering/biomedical design factors:

   a. Operator controls

   b. Operator displays

   c. Maintenance features

   d. Anthropometry

   e. Safety features and emergency equipment

   f. Work space layout

   g. Internal environmental conditions (noise, lighting, ventilation, etc.)

   h. Training equipment

   i. Personnel accommodations

MIL-STD-1521B
APPENDIX E

50.8 <u>System</u> <u>Safety</u>

50.8.1 Review configuration item detail design for compliance to safety design requirements.

50.8.2 Review acceptance test requirements to insure adequate safety requirements are reflected therein.

50.8.3 Evaluate adequacy of detailed design for safety and protective equipment/devices.

50.8.4 Review configuration item operational maintenance safety analyses and procedures.

50.9 <u>Natural</u> <u>Environment</u>

50.9.1 Review detail design to determine that it meets natural environment requirements of the hardware Development Specification.

50.9.2 Insure that studies have been accomplished concerning effects of the natural environment on, or interactions with, the HWCI. Studies which have been in progress shall be complete at this time.

50.9.3 Determine whether arrangements have been made to obtain current and/or forecast natural environment information, when needed for certain HWCIs. Assure compatibility of HWCI and source of information by comparing electrical characteristics and formats for the source and the HWCI.

50.10 <u>Equipment</u> <u>and</u> <u>Parts</u> <u>Standardization.</u>

50.10.1 <u>Equipment</u> <u>and</u> <u>Components.</u> Determine that every reasonable action has been taken to fulfill the standardization requirements for use of standard items (standard item with NSN should be first preference) and to obtain approval for use of non-standard or non-preferred items. Accordingly, the following criteria shall be evaluated:

    a. Data sources that were reviewed.

    b. Factors that were considered in the decision to reject known similar, existing designs.

    c. Factors that were considered in decisions to accept any existing designs which were incorporated, and the trade-offs, if any, that had to be made.

50.10.2 <u>Parts</u>

    a. Determine whether there are any outstanding non-standard or

MIL-STD-1521B
APPENDIX E

      non-preferred parts approval requests and action necessary for approval or disapproval. (Status of parts control program operations).

  b. Identify non-standard-non-preferred parts approval problems and status of actions toward resolving the problems.

  c. Review potential fabrication/production line delays due to non-availability of standard or preferred parts. In such cases, determine whether it is planned to request use of parts which may be replaced by standard items during subsequent support repair cycles. Assure that appropriate documentation makes note of these items and that standard replacement items shall be provisioned for support and used for repair.

  d. Require certification that maximum practical interchangeability of parts exists among components, assemblies, and HWCIs. Reservations concerning interchangeability are identified, particularly for hardness critical items.

  e. Sample preliminary drawings and cross check to insure that parts indicated on the drawings are compatible with the Program Parts Selection List.

50.10.3 <u>Assignment</u> <u>of</u> <u>Official</u> <u>Nomenclature.</u>

  a. Determine whether official nomenclature and approval of nameplates have been obtained to extent practical.

  b. Determine whether DD Form 61, Request for Nomenclature, has been processed to the agreed level of indenture.

  c. Insure that approved nomenclature has been reflected in the Development and Product Specifications.

  d. Identify problems associated with nomenclature requests (DD-61s) together with status of actions towards resolving the problems.

  e. Insure that a software inventory numbering system has been agreed to and implemented to the CSCI level.

50.11 <u>Value Engineering (VE)</u>

50.11.1 Review status of all VECPs presented per the terms of the contract.

50.11.2 Review any new areas of potential Value Engineering considered profitable to challenge.

MIL-STD-1521B
APPENDIX E

50.11.3 If required by contract (funded VE program), review the actual Value Engineering accomplishments against the planned VE program.

50.12 Transportability

50.12.1 Review transportability evaluations accomplished for those items identified as outsized, overweight, sensitive, and/or requiring special temperature and humidity controls.

50.12.2 Review actions taken as a result of the above evaluation to insure adequate facilities and military or commercial transporting equipment are available to support system requirements during Production and Deployment Phases.

50.12.3 Review design of special materials handling equipment, when required, and action taken to acquire equipment.

50.12.4 Insure DOD Certificates of Essentiality for movement of equipment have been obtained for equipment exceeding limitations of criteria established in contract requirements.

50.12.5 Insure transportability approval has been annotated on design documents and shall remain as long as no design changes are made that modify significant transportability parameters.

50.12.6 Identify equipment to be test loaded for air transportability of material in Military Aircraft.

50.13 Test

50.13.1 Review updating changes to all specifications subsequent to the PDR, to determine whether Section 4.0 of the specifications adequately reflects these changes.

50.13.2 Review all available test documentation for currency, technical adequacy, and compatibility with Section 4.0 of all Specification requirements.

50.13.3 For any development model, prototype, etc., on which testing may have been performed, examine test results for design compliance with hardware Development, Software Requirements, and Interface Requirements Specification requirements.

50.13.4 Review quality assurance provisions/qualification requirements in HWCI Product, Software Requirements, or Interface Requirements Specifications for completeness and technical adequacy. Section 4.0 of these specifications shall include the minimum requirements that the item, materiel, or process must meet to be acceptable.

50.13.5 Review all test documentation required to support test

MIL-STD-1521B
APPENDIX E

requirements of Section 4.0 of HWCI Product Specifications for compatibility, technical adequacy, and completeness.

50.13.6 Inspect any breadboards, mockups, or prototype hardware available for test program implications.

50.13.7 Review Software Test Descriptions to ensure they are consistent with the Software Test Plan and they thoroughly identify necessary parameters and prerequisites to enable execution of each planned software test and monitoring of test results. As a minimum, test descriptions shall identify the following for each test:

  a. Required preset hardware and software conditions and the necessary input data, including the source for all data.

  b. Criteria for evaluating test results.

  c. Prerequisite conditions to be established or set prior to test execution.

  d. Expected or predicted test results.

50.14 <u>Maintenance and Maintenance Data</u>

50.14.1 Review adequacy of maintenance plans.

50.14.2 Review status of unresolved maintenance and maintenance data problems since the PDR.

50.14.3 Review status of compliance with Data Item titled "Reliability, Maintainability Data Reporting and Feedback Failure Summary Reports."

50.15 <u>Spare Parts and Government Furnished Property (GFP)</u>.

50.15.1 Review provisioning planning through normal logistics channels and Administrative Contracting Officer (ACO) representative (Industrial Specialist) to insure its compatibility (content and time phasing) with contractual requirements (data and SOW items). The end objective is to provision by a method which shall insure system supportability at operational date of the first site. Also accomplish the following:

  a. Insure contractor understanding of contractual requirements, including time phasing, instructions from logistics support agencies, interim release authority and procedure, and responsibility to deliver spare/repair parts by need date.

  b. Determine that scheduled provisioning actions, such as, guidance meetings, interim release and screening, are being accomplished adequately and on time.

MIL-STD-1521B
APPENDIX E

   c. Identify existing or potential provisioning problems.

50.15.2 Determine quantitative and qualitative adequacy of provisioning drawings and data. Verify that Logistics Critical items are listed for consideration and that adequate procedures exist for reflecting design change information in provisioning documentation and Technical Orders.

50.15.3 Insure support requirements have been determined for installation, checkout, and test for approval by contracting agency. Insure screening has been accomplished and results are included into support requirements lists.

50.15.4 Determine that adequate storage space requirements have been programmed for on-site handling of Installation and Checkout (I&C), test support material, and a scheme has been developed for "down streaming" and joint use of insurance (high cost) or catastrophic failure support items.

50.15.5 Assure that Acquisition Method Coding (AMC) is considered.

50.16 **Packaging/SDPE**

50.16.1 Review proposed package design to insure that adequate protection to the HWCI, and the media on which the CSCI is recorded, is provided against natural and induced environments/hazards to which the equipment will be subjected throughout its life cycle, and to insure compliance with contractual requirements. Such analysis shall include, but not be limited to, the following:

   a. Methods of preservation

   b. Physical/mechanical/shock protection including cushioning media, shock mounting and isolation features, load factors, support pads, cushioning devices, blocking and bracing, etc.

   c. Mounting facilities and securing/hold-down provisions

   d. Interior and exterior container designs.

   e. Handling provisions and compatibility with aircraft materials handling system (463L)

   f. Container marking

   g. Consideration and identification of dangerous/hazardous commodities

50.16.2 Review design of SDPE HWCI to determine if a category I container is required. The analysis of the proposed container or handling, shipping equivalent shall encompass as a minimum:

MIL-STD-1521B
APPENDIX E

a. Location and type of internal mounting or attaching provisions

b. Vibration - shock isolation features, based on the pre-determined fragility rating (or other constraint of the item to be shipped.)

c. Service items (indicators, relief valves, etc.)

d. Environmental control features

e. External handling, stacking and tie-down provisions with stress ratings

f. Dimensional and weight data (gross and net)

g. Bill-of-material

h. Marking provisions including the center-of-gravity location

i. For wheeled SDPE (self-powered or tractor/trailer) the overall length, width, and height with mounted item, turning radius, mobility, number of axles, unit contact load, number of tires, etc.

j. Position and travel of adjustable wheels, titling, or other adjustments to facilitate loading.

50.16.3 Review the results of trade studies, engineering analyses, etc., to substantiate selected package/SDPE design approach, choice of materials, handling provisions, environmental features, etc.

50.16.4 Insure that package/SDPE design provides reasonable balance between cost and desired performance.

50.16.5 Review all preproduction test results of the prototype package design to insure that the HWCI is afforded the proper degree of protection.

50.16.6 Review Section 5, Packaging, of the HWCI Product Specification for correct format, accuracy and technical adequacy.

50.16.7 Review contractor procedures to assure that the requirements of Section 5, Preparation for Delivery of the approved HWCI Product Specification, will be incorporated into the package design data for provisioned spares.

50.17 System Allocation Document

50.17.1 Review maintenance of the System Allocation Document since PDR.

MIL-STD-1521B
APPENDIX E

50.17.2 Insure plans are initiated for configuration item re-allocations that may be necessary due to actions occurring prior to, or during, CDR.

50.18 Design Producibility and Manufacturing

50.18.1 Review the status of all producibility (and productivity) efforts for cost and schedule considerations.

50.18.2 Review the status of efforts to resolve manufacturing concerns identified in previous technical reviews and their cost and schedule impact to the production program.

50.18.3 Review the status of Manufacturing Technology programs and other previously recommended actions to reduce cost, manufacturing risk and industrial base concerns.

50.18.4 Identify open manufacturing concerns that require additional direction/effort to minimize risk to the production program.

50.18.5 Review the status of manufacturing engineering efforts, tooling and test equipment demonstrations, proofing of new materials, processes, methods, and special tooling/test equipment.

50.18.6 Review the intended manufacturing management system and organization for the production program in order to show how their efforts will effect a smooth transition into production.

50.19 Post Review Action

50.19.1 After completing the CDR, the contractor shall publish and distribute copies of Review minutes. The contracting agency officially acknowledges completion of a CDR as indicated in paragraph 4.2.4.

50.19.2 The accomplishment of the CDR shall be recorded on the configuration item Development Record by the contractor.

Index

A
actual cost of work performed (ACWP), 227
allocated requirements, 13, 42-43
allocation of resources, 235
  electromagnetic compatibility, 137
  human engineering, 114
  maintainability of design, 103
  reliability of design, 84
  requirements allocation sheet (RAS), 45
  system engineering management plan (SEMP), 235
alternatives analysis (*see also* systems analysis), 31
analogous cost method, 60
apportioned effort, 216
audits, 201-205, 269
  auditor responsibility, 201-202
  measurements, 203-204
  performing an audit, 202-203
  reporting results, 204-205

B
bar charts, 223-224
baselining, 42, 193, 256, 262
Boothroyd-Dewhurst Design for Assembly software, 169
Brooks, Frederick, 160
budget at completion (BAC), 227
budgeted cost for work performed (BCWP), 227, 228
budgeted cost for work scheduled (BCWS), 227, 228
budgeting, 226
built-in tests, 147
burn-in testing, 53

C
capacitive coupling, EMC, 132
change control board (CCB), 194
changes management, 46-47, 192-194
  safety considerations, 129
communication, 6
complexity-related issues, 6
computer-resources support, 69
conducted emissions and susceptibility, EMC, 132
configuration management, 192-194
  baselines, 193
  change control board (CCB), 194
  changes management, 192-194
  interfaces, 193
constraints, 9
  maintainability of design, 101

problem-solving, 14, 18-19
system engineering management plan (SEMP), 234-235
systems analysis, 33
value system design, 23
Consumer Product Safety Act, 120
contract work breakdown structure (CWBS), 261
contractors/suppliers, 258
  evaluation, 248-249
  maintainability of design, 102
  reliability of design, 83
  reviewing, 200-201
  safety considerations, 124
controlled convergence method, systems analysis, 33-35
cost control, 221-222, 226-229, 256, 259, 266
  actual cost of work performed (ACWP), 227
  budget at completion (BAC), 227
  budgeted cost for work performed (BCWP), 227, 228
  budgeted cost for work scheduled (BCWS), 227, 228
  complexity-related issues, 225
  control techniques, 221-222
  CPM charts, 223-224
  design to cost method, 177-179
  earned value, 227
  effective scheduling, 224
  estimate at completion (EAC), 227
  estimating cost and time at completion, 229
  Gantt or bar charts, 223-224
  life cycle costs, 59-66, 266
  milestone charts, 223-224
  PERT charts, 223-224
  progress measurement by schedule, 224-225
  software design, 160-162
  task timing, 225-226
Cost Estimating Relationships (CER), 60
CPM charts, 223-224
creativity, innovation, invention, 10
crew maintenance, 100
criteria, 23, 33
critical design review (CDR), 263, 270, 304-319
critical items, reliability of design, 86
critical tasks, 115
cross-correlation charts, 20-21
customer satisfaction, 19, 79-80

D

data collection/analysis, 103-105, 266
  maintainability of design, 104-105
decision-making, 6, 41, 257
decomposition, 12, 42-45
  allocation, 42-43
  interface identification, 44
  Pareto rule, 44
  Quality Function Deployment (QFD), 43-44
degree-of-need identification, 15
depot maintenance, 99-100
derating techniques, 88
derived requirements, 13
description/documentation, 12, 45-46, 256, 260, 265
  requirements allocation sheet (RAS), 45
design, 7, 256
  Design for Assembly (DFA), 169
  Design for Manufacture (DFM), 166-167
  design reviews, 199, 263
  design to cost methods, 177-179
  electromagnetic compatibility, 136-139
  integrated approach to system design, 7
  Joint Application Design (JAD), 19-20
  maintainability of design, 96, 105, 106-108
  producibility/manufacturability, 170, 171
  reliability of design, 78, 88-91
  safety considerations, 128
  supportability of design, 71
  test results and design, 54
  testability of design, 152, 153-154
Design for Assembly (DFA), 169
Design for Manufacture (DFM), 166-167
design reviews, 199, 263
design to cost method, 177-179
destructive testing, 53
dictionary, work breakdown structure, 209, 211, 213
direct support, 100
discrete effort, 216
documentation (*see* description/documentation)
DOD-HDBK-743 Anthropometry (*see* human engineering)
dynamic simulation, human engineering, 117

320

E

effectiveness analysis modeling, 266
electromagnetic compatibility, 131-146
  allocation of requirements, 137
  analysis of EMC, 136
  benefits derived, 133-134
  capacitive coupling, 132
  conducted emissions and susceptibility, 132
  control plan, 136controlling EMC/Control Plan, 139-142
  defining EMC, 131-133
  design reviews, 138
  "designing-in" EMC, 136-137, 138
  electromagnetic signals, 132
  electrostatic discharge (ESD), 132
  emission level-settings, 134-135
  environment defined for EMC, 135-136
  interfaces, 189
  integrating system parameters, 137
  interference sources, man-made, 133
  predicting EMC, 138
  problem area identification, 136
  radiated emissions and susceptibility, 132
  requirements analysis, 134
  risk reduction steps, 136
  safety considerations, 134
  standardization of design, 137-138
  static electricity, 132
  task breakdown, 134-138
  Test Plan, 142-145
  verification of design, 138
electrostatic discharge (ESD), 132
engineering cost method, 60
engineering specialty integration, 239, 255, 257, 258
environmental profile charts, 16-17
environmental stress screening (ESS), reliability of design, 87
environmental testing, 53
environments, 3-4
  electromagnetic compatibility, 135-136
  human engineering, 117
  problem-solving, 15-16
equipment
  design, human engineering, 116
  selection, human engineering, 115
estimate at completion (EAC), 227
estimating, work breakdown structure, 219

F

facilities, 69, 117
failure modes and effects analysis (FMEA), 104
failure reporting/analysis/corrective action system (FRACAS), 84
failures of system, 77
  cost of failure, 80
  derating techniques, 88
feasibility studies, 32
Fisher, R.A., 37
formal qualification review (FQR), 270
functional analysis, 11, 25, 26-31, 264
  flow chart, functional, 27-29
  human engineering, 112-113
  identifying functions, 26
  indenture levels, 28
  limitations, 30-31
  time-critical functions, 30
  time-line analysis, 30
  traceability, 28
  value analysis use, 175
functional configuration audit (FCA), 270
functional flow chart, 27-29

G

Gantt charts, 223-224
go-no go testing, 53

H

Hall, Arthur D., 10
hierarchical level testing, 53
Hill, J. Douglas, 10, 43
human engineering, 111-119
  allocation of system functions, 114, 115
  controlling human engineering, 118
  critical tasks defined, 115
  documentation associated with human engineering, 113
  DOD-HDBK-743 Anthropometry, 113
  dynamic simulation, 117
  equipment detail design, 116
  equipment selection, 115
  facilities design, 117
  functional analysis, 112-113
  human error, 112
  human factors, 111
  information flow and processing, 114
  MIL-H-46855B Human Engineering, 111, 113
  MIL-HDBK-759 Human Factors, 113
  MIL-STD-1472 Human Engineering, 113
  mockups, 116
  modeling, 116
  personnel and staffing, 112
  potential operator capabilities, 114
  preliminary design, 116
  procedure development, 117
  requirements analysis, 112-114
  SAE J925 Minimum Access Dimensions, 113
  studies, experiments, tests, 116
  task analysis, 111, 115
  task breakdown, 114-117
  testing, 117
  time-line analysis, 113
  training, 112
  work environment, 117
human error, 112

I

IEEE Std 830-1984 Good Software Design, 157
indenture levels, functional analysis, 28
information, 10
inspections, software design, 162
integrated logistic support (ILS), 259-260
  supportability of design, 67, 70-71
interaction analysis, 16
interfaces, 44, 69, 189-190, 257
  configuration management, 193
  safety considerations, 129
interference sources, man-made (EMC), 133
intermediate maintenance, 99
iteration, 9

J

Joint Application Design (JAD), 18-20

L

Lano, R.J., 20
leadership concepts, 181-182
liability, legal, safety considerations, 121
life cycle costs, 59-66
  analogous cost method, 60
  analysis procedures, 64
  benefits derived, 63
  breakdown of costs, 60-62
  controlling life cycle costs, 65-66
  Cost Estimating Relationships (CER), 60
  engineering cost method, 60
  input data for life cycle costs modeling, 62-63
  modeling, 62
  parameters and factors influencing life cycle costs, 59
  parametric cost method, 60
  reliability of design, 63, 80
  system support, 63

Index

task breakdown, 64-65
trade-off analysis, 63-64
logic of systems engineering, 10-12
logistic engineering, 266
logistic support analysis (LSA), 105,
loss calculations, value system design, 24-25

M

maintainability of design, 69, 95-110
  allocations, 103
  availability and dependability factors, 97
  benefits derived, 96-97
  constraints, 101
  contractors/vendors, 102
  controlling maintainability, 108-109
  crew maintenance, 100
  data collection/analysis, 103
  demonstrations, 105-106
  depot maintenance, 99
  design criteria, 105
  "designing-in" maintainability 96, 106-108
  direct support, 100
  elements of maintainability program, 96-97
  failure modes effects analysis (FMEA), 104
  general support, 100
  interfaces, 189
  intermediate maintenance, 99
  logistics support analysis (LSA), 105
  Maintainability Program Plan, 108-109
  maintenance concepts, 98
  maintenance plans, 98
  mean-time-to-repair (MTTR), 95
  MIL-HDBK-338 Maintenance, 98
  MIL-STD-470A Maintainability Program, 101-106, 108-109
  MIL-STD-721C Reliability/Maintainability, 95
  military maintenance level standards, 98-100
  modeling, 103
  organizational maintenance, 98
  planning, 105
  predictions, 104
  preliminary design review (PDR), 103
  qualitative requirements, 101
  quantitative requirements, 101
  reliability of design, 86
  requirements analysis, 97-101
  responsibility flowchart, 100
  reviews, 102
software design, 160
task breakdown, 101-106
maintenance engineering analysis (MEA), 265
mean-time-between-failure (MTBF), 77, 81-82
mean-time-to-repair (MTTR), 95
MIL-H-46855B Human Engineering (*see* human engineering)
MIL-HDBK-338 Maintenance (*see* maintainability of design)
MIL-HDBK-759 Human Factors (*see* human engineering)
MIL-S-83490 Specifications, 267
MIL-STD-1309C Test Measurements/Diagnostics, 147
MIL-STD-1388-1A Logistic Support Analysis, 72-73
MIL-STD-1472 Human Engineering (*see* human engineering)
MIL-STD-1521 Technical Reviews and Audits, 197, 262, 268-319
MIL-STD-2165 Testability Program (*see* testability of design
MIL-STD-2167 Software Design (*see* software design)
MIL-STD-470A Maintainability Program (*see* maintainability)
MIL-STD-480 Configuration Control, 254
MIL-STD-483 Configuration Management, 255, 257
MIL-STD-490 Specifications (*see* specifications)
MIL-STD-499A (*see* system engineering management plan (SEMP)
MIL-STD-721C Reliability/Maintainability (*see* maintainability)
MIL-STD-785B Reliability Program (*see* reliability of design)
MIL-STD-83490, 267
MIL-STD-881 Work Breakdown Structures, 255-256
MIL-STD-882B System Safety (*see* safety considerations)
Miles, Lawrence, 173
milestone charts, 223-224
mission profiles, 14
mockups, human engineering, 116
modeling
  effective analysis, 256, 266
  human engineering, 116
  life cycle costs, 62
  logistics support, 266
  maintainability of design, 103
reliability of design, 84
systems analysis, 37
verification methods, 50
Morris, Chuck, 19
Mythical Man-Month, The, 160

N

N-squared charts, 20-21
National Electrical Code (NEC), 122
new product/design management, 7
normal vs. out-of-normal conditions, problem-solving, 18

O

objectives, 256
  constraints, 23
  criteria, 23
  measurement of objectives, 22-23
  systems engineering, 5
  value system design, 22
  weighting factors, 23-24
Occupational Safety and Health Administration (OSHA), 121-122
optimization, 35-37, 266
organizational maintenance, 98

P

packaging, 69, 86
parameters, TPM, 259, 262
parametric cost method, 60
Pareto rule, 44, 178
partitioning of problems, 15
partitioning, interfaces, 189
personnel, 69
  human engineering, 112
  key-people identification, 15
  safety considerations, qualifications, 124-125
PERT charts, 223-224
physical configuration audit (PCA), 270
planning
  system engineering management, 233-234
  TPM, 262
predictions
  electromagnetic compatibility, 138
  maintainability of design, 104
  reliability of design, 85
preliminary design review (PDR), 263, 269, 284-303
  maintainability of design, 103
  reliability of design 83
priority analysis, 32
problem-solving, 6, 11, 12-22
  alternative analysis, 15
  bounding of problems, 15
  constraints, 14, 18-19
  cross-correlation charts, 20-21

Index   323

customer identification, 19
degree-of-need identification, 15
environments, 15-16
interaction analysis, 16
Joint Application Design (JAD), 18-20
key-people identification, 15
N-squared charts, 20-21
noncustomer interactive analysis, 20
normal vs. out-of-normal conditions, 18
partitioning of problems, 15
Quality Function Deployment (QFD), 19-21
requirements analysis, 12-14, 18-19
reviews, 199
scenarios, 14-15
scoping problems, 15
self-interaction matrix, 20-21
tools, 20
procurement, specification-writing, 188-189
producibility/manufacturability, 165-172
 benefits derived, 167-168
 design factors, 170, 171
 Design for Assembly (DFA), 169
 Design For Manufacture (DFM), 166-167
 engineering/manufacturing interface, 167
 essential producibility, 172
 implementing producibility/manufacturability, 168
 process-capability analysis, 170-171
 risk assessment, 171
 task breakdown, 169
product design, verification methods, 51-53
production assessment testing, 53
production engineering analysis, 267
production reliability acceptance test (PRAT), 87
proposals, 243-249
 content of proposal, 244-245
 evaluating proposals, 246-249
 evaluators, writing proposals for, 245-246
 planning sections, 245
 qualitative standards, 247
 quantitative standards, 247
 risk assessment, 248
 scoring proposals, 246-248
 supplier evaluation, 248-249
 training in art of proposal-writing, 243

value system creation, 246
writing proposals, 244-246
Pugh controlled convergence method, systems analysis, 33-35

Q

qualitative requirements, 13, 82, 101
Quality Function Deployment (QFD), 19, 21, 43-44
quality loss function, 39-41
quantitative requirements, 13, 81, 101

R

radiated emissions and susceptibility, EMC, 132
reliability development/growth test (RDGT) program, 87
reliability of design, 63, 76-94
 allocations, 84
 benefits derived, 78-80
 contractors/vendors/suppliers, 83
 controlling reliability, 91-93
 cost of failures, 80
 critical item identification, 86
 customer satisfaction, 79-80
 derating systems, 88
 "designing-in" reliability, 78, 88-91
 electronics design checklist, example, 89-91
 environmental stress screening (ESS), 87
 failure modes effects criticality analysis (FMECA), 85
 failure of systems, 77
 failure reporting analysis corrective action system (FRACAS), 84
 failures of system, 88
 improvements to other systems/system areas, 78-80
 interfaces, 189
 life cycle costs, 80
 mean-time-between-failure (MTBF), 77, 81-82
 MIL-STD-721C Reliability/Maintainability, 76, 83-88
 MIL-STD-785B Reliability Program, 91-93
 modeling, 84
 predictions, 85
 preliminary design review, 83
 production reliability acceptance test (PRAT), 87
 qualitative requirements, 82
 quantified requirements, 81
 reliability development/growth test (RDGT) program, 87

reliability qualification test (RQT) program, 87
 requirements analysis, 80-82
 reviews, 83
 safety considerations, 81
 sneak circuit analysis (SCA), 85
 software design, 159
 standardization of parts, 86
 task breakdown, 82-88
 testing, storage, handling, packaging, transport, maintenance, 86
 tolerance analysis, 85-86
 warranties, 80
reliability qualification test (RQT), 87
repair levels, 266
requirements allocation sheet (RAS), 45
requirements analysis, 4, 264
 allocated requirements, 13
 derived requirements, 13
 electromagnetic compatibility, 134
 human engineering, 112-114
 maintainability of design, 97-101
 problem-solving, 12-13, 18-19
 qualitative requirements, 13, 82, 101
 quantitative requirements, 13, 81, 101
 reliability of design, 80-82
 requirements allocation sheet (RAS), 45
 safety considerations, 122-123
 software design, 157-160
 specifications, 184
 stated requirements, 14
 testability of design, 149-152
responsibility assignment matrix (RAM), 211
reviews, 191, 197-201, 262-263, 269
 conducting reviews, 200
 design reviews, 199
 goal setting, 197
 maintainability of design, 102
 MIL-STD-1521B Technical Reviews/Audits, 197, 262-263, 268-319
 objectives, 197
 problem resolution, 199
 reliability of design, 83
 safety considerations, 124
 suppliers, reviewing suppliers, 200-201
 walk-throughs, 199
risk assessment/management, 6, 195-196, 198, 261
 producibility/manufacturability, 171
 proposals, 248

Index

safety considerations, 120
software design, 163
systems analysis, 33, 35

S

SAE J925 Minimum Access Dimensions (*see* human engineering)
safety considerations, 120-130
  assessment of safety procedures, 127
  benefits derived, 120-121
  code-level software hazards, 128-129
  compliance assessment, 127
  Consumer Product Safety Act, 120
  contractors/vendors/suppliers, 124
  controlling safety, 130
  detailed design hazards, 128
  deviations/waivers, 127
  electrical systems, example, 123
  electromagnetic compatibility, 134
  GFE/GFP system safety analysis, 127
  interface analysis, software/user, 129
  interfaces, 189
  liabilities, legal, 121
  MIL-STD-882B System Safety, 120, 123-129
  National Electrical Code (NEC), 122
  occupational health hazard assessment, 126
  Occupational Safety and Health Administration (OSHA), 121-122
  preliminary hazard list/analysis, 125
  progress summary, 124
  qualifications of key personnel, 124-125
  reliability of design, 81
  requirements analysis, 122-123
  risk analysis, 120
  software change hazards, 129
  software requirements hazard analysis, 127
  software testing, 129
  subsystem hazard, 125
  support, 124, 126
  system hazards, 126
  task breakdown, 123-129
  testing, 124
  top-level design hazards, 128
  tracking hazards, 124
  training, 127
  U.S. Safety Laws, 121
  verification, 127
  workplace standards, 121
Sage, Andrew, 10
scheduling control, 221-226, 259
  software design, 160-162
scoping techniques, 15
self-interaction matrix, 20-21
sensitivity analysis, 41
signal-to-noise ratio, 39-41
simulation
  dynamic simulation, human engineering, 117
  systems analysis, 37
  verification methods, 50
sneak circuit analysis (SCA), reliability of design, 85
software design, 157-164
  attributes of good software, 159
  cost and time estimates, 160-162
  delegating control of software, 163
  design discipline, MIL-STD-2167, 162
  flow-down of requirements, 158-159
  IEEE Std 830-1984 Good Software Design, 157
  inspections, 162
  maintainability of design, 160
  reliability of design, 159
  requirements analysis, 157-160
  risk management, 163
  task definitions, 160-161
  technical control, 162-163
  technical planning steps, 160-162
  verification, 163
software specification review (SSR), 269, 282-283
specialty engineering, 57
specification trees, 186, 261
specifications, 183-189, 267
  commercial type, 188
  hierarchy of specifications, 186, 188
  MIL-STD-490 Specifications, 185
  procurement, writing specs, 188-189
  requirements analysis, 184
  specification sections and appendix, 185-187
  statement of work (SOW), 190
  writing process, 186
standardization
  electromagnetic compatibility designs, 137-138
  reliability of design, 86
statement of work (SOW), 190-193
  descriptions, 190
  preparation of SOW, 190-192
  specifications, 190
static electricity, EMC, 132
statistical analysis, 37-39
storage, 69-86
suppliers (*see* contractors/vendors/suppliers)
supply support, 69, (*see also* logistics engineering and maintainability)
support equipment/systems, 69
supportability of design, 67-75
  benefits derived, 72
  computer-resources support, 69
  controlling supportability, 73-74
  design factors, 71
  engineering activities related to, 72
  facilities, 69
  integrated logistics support (ILS) concepts, 67, 70-71
  interface design, 69
  issues and considerations, 68-69
  logistics elements, example, 70
  maintenance planning, 69
  MIL-STD-1388-1A, 72-73
  packaging, handling, storage, transportability, 69
  personnel requirements, 69
  supply support, 69
  support equipment, 69
  task breakdown, 72-73
  technical data, 69
  training, 69
  transportability, example, 71
synthesis, 12, 31, 265, 271, 275
system design review (SDR), 263, 269, 274-281
System Engineering Management Plan (SEMP), 54, 232-242, 257
  constraints, 234-235
  effectiveness measurements, 236
  engineering specialty integration, 239, 255
  general criteria for engineering management, 256, 257
  improving system engineering process, 236, 255, 264-267
  MIL-STD-499A, 237-242
  outline example, 239-241
  planning and decision-making, 233-234
  relating SEMP to other plans, 235
  resource allocation, 235
  system engineering process defined, 238
  tailoring SEMP, 234, 242, 254
  technical performance measurement (TPM), 195
  technical program planning and control, 238, 255, 261-264
  time-phasing SEMP, 234
system requirements review (SRR), 263, 269, 271-273
system support, 63
systems analysis, 12, 31-42
  baselining, 42
  constraints, 33
  criteria, 33
  decision-making, 41
  feasibility studies, 32
  modeling, 37

optimization, 35-37
priority analysis, 32
Pugh controlled convergence method, 33-35
risk analysis, 33, 35
sensitivity analysis, 41
simulation, 37
statistical analysis, 37-39
Taguchi quality-analysis methods, 39-41
trade studies, 41-42
utilities, 24-25
weighting factors, 33

T

task analysis, human engineering, 111, 115
task timing, cost control, 225-226
technical control, 183-207
  auditing systems engineering, 201-205
  configuration management, 192-194
  interfaces, 189-190
  reviews, 197-201
  risk management, 195-196, 198
  specifications, 183-189
  statement of work (SOW), 190-192
  technical performance measurement (TPM), 194-195
technical performance measurement (TPM), 194-195, 256, 259, 261-262, 272, 276
technical program planning and control, 255, 257-258
technical reviews, 197-201, 262-263, 269-319
Test and Evaluation Master Plan (TEMP), 54-55
test readiness review (TRR), 270
testability of design, 147-156
  benefits derived, 148
  built-in testing, 147
  controlling testability, 154-155
  data collection/analysis, 151
  demonstrating testability, 152
  design analysis/design reviews, 152
  "designing-in" testability, 153-154
  interfaces, 189
  MIL-STD-1309C Test Measurements/Diagnostics, 147
  MIL-STD-2165 Testability Program, 150-153
  objectives of testability program, 149
  planning, 150
  requirements analysis, 149-152
  reviews, 150-151
  task breakdown, 150-153
  tasks description, input, output, 155
  testability checklist, partial, 153-154
  typical testability values, 149
testing (*see also* verification methods), 52-55, 261
  built-in testing, 147
  design vs. test results, 54
  human engineering, 117
  reducing integration and test time, 55
  reliability of design, 86
  safety considerations, 124, 129
  Test and Evaluation Master Plan (TEMP), 54-55
  testability of design factors, 147-156
time-critical functions, 30
time-line analysis, 30
  human engineering, 113
time-phasing, 10-11, 214, 234
tolerances, reliability of design, 85-86
traceability, functional analysis, 28
trade studies, systems analysis, 41-42
trade-off analysis, 63-64, 266
training, 69
  human engineering, 112
  proposal-preparation, 243
  safety considerations, 127
transportability, 69, 71, 86

U

utilities
  systems analysis, 33
  value system design, 24-25

V

VA/VE (*see* value analysis)
validation, verification methods vs., 50
value analysis, 173-176
  benefits derived, 174
  checklist, 175-176
  functional analysis role, 175
  implementation steps, 174
  team assembled for value analysis, 174-175
value engineering (*see* value analysis)
value system design, 11, 22-25, 246
  constraints, 23
  criteria, 23
  loss calculations, 24-25
  objectives and objectives-measurement, 22-23
  utilities, 24-25
  weighting factors, 23-24
variables testing, 53
vendors (*see* contractors/suppliers)
verification methods (*see also* testing), 50-56
  controlling verification steps, 52
  electromagnetic compatibility, 138
  modeling, 50
  planning for verification, 52
  product verification techniques, 51-53
  safety considerations, 127
  simulation, 50
  software design, 163
  testing, 52-55
  validation vs., 50

W

walk-through reviews, 199
Warfield, John, 10, 43
warranties, 80
weighting factors, 23-24, 33
work authorizations, 240, 264
work breakdown structure, 208-220, 261
  apportioned effort, 216
  benefits derived, 210-211
  criteria for inclusion, 216
  development steps, 211, 214-219
  dictionary, 208-209, 211, 213
  discrete effort, 216
  elements, 208, 210, 213, 214
  estimations, 219
  garage door opener, example, 210
  integration of organization structure and WBS, 212
  level of effort, 216
  levels, 214
  responsibility assignment matrix (RAM), 211
  revising WBS, 219
  system-level efforts, 218
  time phasing, 214
  tree format, 208-209
  work package budgets, 209
  work packages, 209, 216-218
work packages, work breakdown structure, 209, 216-218

About the Author

Jim Lacy trains and consults in systems engineering and project management. He teaches graduate level engineering for Texas Tech University and Southern Methodist University.

Jim has worked in the positions of program manager, systems engineer, and design engineer. His engineering experience includes all phases of the product life cycle from research and development, proposals, design, testing, production, to field service.

He has served on two national standards committees and as a Director for a national council on systems engineering. He is a Registered Professional Engineer in the state of Texas.

Jim Lacy receives business mail at P.O. Box 38183, Dallas, TX 75238.

Order Form

Jim Lacy Consulting
PO Box 38183
Dallas, TX 75238 USA
Telephone (214) 341-0022

Please send me my own copy of *Systems Engineering Management: Achieving Total Quality*.

_____ copies @ $70.00 each   **Texas addresses** $75.78 each, includes sales tax.

Send check or money order. UPS shipping is included.

Name:_____

Address:_____

_____

_____ZIP_____

Phone: (_____) _____